Trilobites of the Tremadoc Bjørkåsholmen Formation in the Oslo Region, Norway

JAN OVE RØYSI EBBESTAD

Ebbestad, J.O.R. 1999 11 15: Trilobites of the Tremadoc Bjørkåsholmen Formation in the Oslo Region, Norway. *Fossils and Strata*, No. 47, pp. 1–118. Oslo. ISSN 0300-9491. ISBN 82-00-37702-4.

The trilobite fauna of the upper Tremadoc Bjørkåsholmen Formation in the Oslo Region is revised and redescribed, recognizing 36 species assigned to 28 genera. The regional and vertical distribution of the trilobite fauna are also discussed. The Bjørkåsholmen Formation is found in the Lower Allochthon of the Synfjell Nappe and across the Oslo Region in Norway. In Sweden it crops out in Västergötland, Scania and on Öland. Generally the unit is 60–120 cm thick, comprising micritic to intrasparitic limestone composed of several individual limestone beds, covering the Zone of *Apatokephalus serratus*. Near the base of the formation a horizon consisting of dark limestone nodules appears, containing a trilobite fauna dominated by *Bienvillia angelini*. This level is found throughout the studied area and is an important correlatable horizon. New material of rare species, such as *Peltocare modestum* Henningsmoen, 1959, *Parabolinella lata* Henningsmoen, 1957, *Falanaspis aliena* Tjernvik, 1956, *Harpides rugosus* (Sars & Boeck, 1838) and *Parapilekia speciosa* (Dalman, 1827), is described and figured. *Orometopus elatifrons* (Angelin, 1854) is recognized as distinctly different from British material formerly assigned to this taxon. Three new species are described, *Saltaspis stenolimbatus* n.sp., *Apatokephalus dactylotypos* n.sp., and *Niobe* (*Niobella*) *eudelopleura* n.sp. Additionally, the species *Apatokephalus* cf. *sarculum* Fortey & Owens 1991 from the upper part of the Alum Shale Formation is described. Biostratigraphical studies were carried out at six localities distributed across the Oslo Region. A relative-abundance distribution shows that *Ceratopyge acicularis* dominates the lower limestone beds above the dark limestone nodules and is followed by a small acme of *Apatokephalus serratus*, then a dominance of *Euloma ornatum*, and, finally, *Symphysurus angustatus* in the uppermost fossiliferous beds of the formation. Species of the large asaphid *Niobe* are present throughout the unit in relatively constant numbers. The remaining species are present in limited numbers. Older views claiming a greater diversity in the Oslo–Asker district compared to the rest of the Oslo Region are erroneous. All data suggest a coherent distribution and diversity across the Oslo Region, with local variations. □*Ceratopyge Limestone, Tremadoc, Ordovician, Norway, Trilobita, systematics.*

Jan Ove Røysi Ebbestad, Paleontological Museum, University of Oslo [present address: Department of Earth Sciences (Historical Geology and Palaeontology), Uppsala University, Norbyvägen 22, SE-752 36 Uppsala, Sweden]; 2nd May, 1996; revised 15th June, 1998.

Contents

Introduction

The Bjørkåsholmen Formation (Norwegian *Bjørkåsholm-formasjonen*), formerly the Ceratopyge Limestone (3aγ), is present throughout most of the Baltic Shield, with equivalents occurring within the Parautochthon and Lower Allochthon of the Scandinavian Caledonides. The unit is characterized by its limited thickness, rich shelly fauna dominated by trilobites, and its broad regional distribution. It was deposited in the shallow epicontinental seas across the Baltic Platform in the Early Ordovician, Tremadoc.

Based on palaeomagnetic data from 75 sites in Southern Scandinavia, Perroud *et al.* (1992) suggested that the Baltic Platform was situated between 50°S and 30°S during the Early Palaeozoic (Fig. 1). Other palaeomagnetic data support this together with sedimentological and palaeontological data (Spjeldnæs 1961; Noltimer & Bergström 1977; Torsvik *et al.* 1990). A rapid (4–7 cm/yr) northwards drift of the Baltic Platform from a temperate to tropical–subtropical position has also been suggested (Jaanusson 1973; Lindström 1984; Torsvik *et al.* 1990; Perroud *et al.* 1992). The drift was accompanied by a large-scale counterclockwise rotation of Baltica, with an estimated 20°/10 Ma rotation speed or shift in position of 110° between the Early and Late Ordovician (Torsvik *et al.* 1990; Perroud *et al.* 1992).

The Baltic Platform extends from the Russian Platform in the east (Størmer 1967) to 200–400 km west of the present Norwegian coastline (Nystuen 1981; Nickelsen *et al.* 1985; Bruton *et al.* 1989) and is divided into three structural units (Fig. 2). The stable platform, constituting autochthonous deposits, extends westwards from Estonia to Västergötland and northwards from Bornholm to Digermulen (Finnmark), an area known as Baltoscandia (Martinsson 1974). The deposits consist of mainly horizontal carbonate sediments (Männil 1966; Størmer 1967). The Ordovician succession is usually less than 200 m thick, representing a deposition rate of 2–3 mm/1000 years (Lindström 1971; Jaanusson 1976). The Middle Ordovician strata are developed in distinct confacies belts, defined as a combination of litho- and biofacial characters that maintained a fairly constant relative position within the depositional area through time (Jaanusson 1976). The Bjørkåsholmen Formation is found in Västergötland and on Öland, belonging to the Lower Ordovician Hunneberg–Modum subconfacies belt (Erdtmann & Paalits 1995). The second structural unit is the foreland of Oslo–Scania–Lysogor, consisting of autochthonous and parautochthonous deposits. The area comprises the Scania confacies belt and four Oslo belts with subconfacies belts (Erdtmann 1965b; Jaanusson 1976; Bockelie 1978; Erdtmann & Paalits 1995). The confacies belts apply to the Upper Cambrian to Middle Ordovician strata (Erdtmann & Paalits 1995). The sediments are dark shales and mudstones with interfingering limestone units. In Scania and the Southern part of the Oslo Region the strata are almost horizontal, while they are folded, cleaved and thrust in the remainder of the Oslo Region. In Scania the Ordovician deposits are 30–200 m thick, increasing to 250–1500 m in the Oslo Region (Størmer 1967; Bruton *et al.* 1985; Bockelie & Nystuen 1985). The Bjørkåsholmen Formation is present throughout the foreland area. The last structural area is the Scandinavian Caledonides, consisting of allochthonous deposits west of the 1800 km long Caledonian front (Bruton *et al.* 1985). The formation of the Caledonides took place during the Silurian and Devonian, owing to east–west convergent movements between Baltica and Laurentia (Ziegler 1982). The Ordovician deposits are preserved as limestones, shales, sandstones, greywackes and crystalline rocks in several nappe units (Bruton *et al.* 1989). Analyses show that some of these sediments must have been deposited 200–400 km west of the present coastline (Nystuen 1981; Nickelsen *et al.* 1985; Bruton *et al.* 1989). An equivalent of the Bjørkåsholmen Formation is found in the Lower Allochthon Synfjell Nappe (Bruton *et al.* 1989).

Owen *et al.* (1990) formally defined the formation based on the hypostratotype at Bjørkåsholmen in Slemmestad, Norway. The base is drawn at the abrupt change

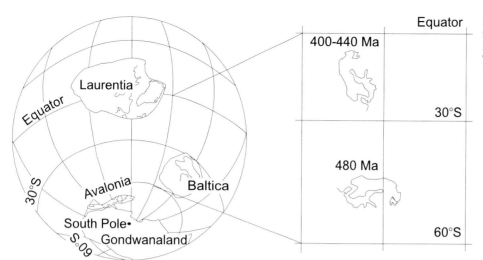

Fig. 1. Palaeogeographic position of the Baltic Shield in the Early and Late Ordovician, respectively (from Torsvik *et al.* 1990; Perroud *et al.* 1992).

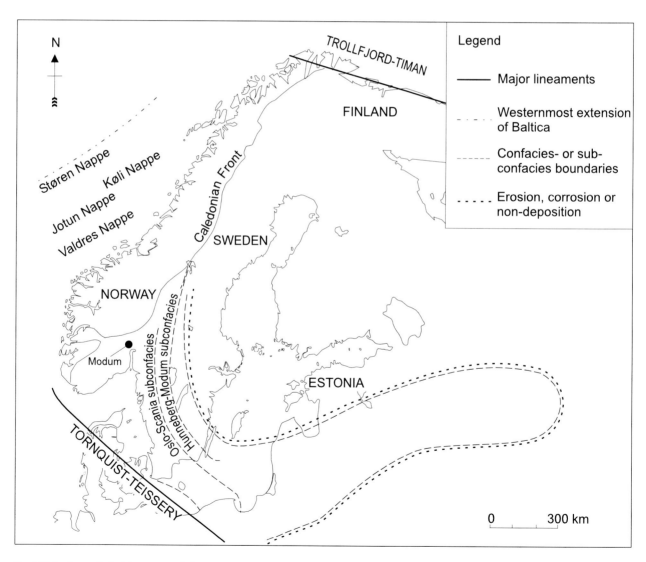

Fig. 2. Major Lower Ordovician structural elements and confacies belts of western and central Baltica (from Bruton & Harper 1985; Bruton *et al.* 1989; Erdtmann & Paalits 1995).

from the dark shale of the underlying Alum Shale Formation. The following limestone unit is generally divided into a lower light-grey micritic limestone and an upper micrite or fine- to medium-grained intrasparite. The topmost beds in the Vestfossen and Oslo–Asker districts contain glauconite-like grains, but this is not recognized further north in the Oslo Region. The thickness of the unit varies between 0.6 and 1.2 m.

The varied and rich trilobite fauna comprises 36 species assigned to 28 genera. Usually only exuviae are preserved, articulated specimens being exceedingly rare. However, the state of preservation is good and the exoskeletal parts are rarely broken, even though they appear in a well-homogenized sediment. The richness and preservation of the fauna in this thin unit is by itself important, but also the position and occurrence of the fauna must be considered. It is not only central in a Baltoscandian context but has general implications for most Lower Ordovician trilobite communities. Most of the genera are widespread, being found in most of the Ordovician continents, and have a short stratigraphical range.

A number of the species represent type species (e.g., *Tropidopyge broeggeri*, *Apatokephalus serratus*, *Ceratopyge acicularis*, *Dikelokephalina dicraeura*, *Orometopus elatifrons*, *Pagometopus gibbus*, *Falanaspis aliena*, *Parapilekia speciosa*), and some inconcistencies in their definitions have been revealed.

There are a few major modern studies discussing aspects of this formation, notably by Tjernvik (1956a), Henningsmoen (1959), Bjørlykke (1974), Bruton *et al.* (1989) and Owen *et al.* (1990). Most of the literature is considerably older. However, several unpublished theses exist, and data from these are incorporated here. These works have usually been widely cited in the literature, but they remain unpublished. The main object of the study presented here is to give a complete revision of the known trilobite fauna of the Bjørkåsholmen Formation in the Oslo Region, Norway. However, during the study it became evident that most of the earlier concepts on the regional and vertical distribution of the formation and its fauna were incorrect or incomplete. It was therefore necessary to include a short bio- and lithostratigraphical discussion.

Historical notes

Most of the knowledge on the Bjørkåsholmen Formation is based on historically early studies briefly reviewed here (Fig. 3). The history of the formation is also closely tied to the development of the *etasje* system in Norway; Swedish stratigraphy developed separately. The older term Ceratopyge Limestone is used in this historical summary rather than the more recent synonym Bjørkåsholmen Forma-

tion. In the remainder of this work however, the modern synonym is consequently referred.

Several trilobites of the Ceratopyge Limestone were among the first to be described in Scandinavia (Dalman 1827; Sars 1835; Boeck 1838; Angelin 1851, 1854). Angelin (1854) introduced *Ceratopyge* in a combined bio- and lithostratigraphical context. His Region 4, *Regio Ceratopygarum BC*, was based on strata in Oslo, Norway, and Hunneberg in Västergötland, Sweden.

A stratigraphical division of the Norwegian strata in the Oslo Region was presented by Kjerulf (1857), with a contribution by Dahll (1857). Lithological units were here named and numbered as *etasjer* in a consecutive manner. The Ceratopyge Limestone with its typical trilobites was tentatively placed within *etasje* 3 (Kjerulf 1857, p. 93). Subsequently the trilobites of the Ceratopyge Limestone and the Ceratopyge Shale were assigned to *etasje* 2 (Kjerulf 1865, pp. 1–3).

In Sweden, Linnarsson (1872, p. 46) correlated the Ceratopyge Limestone of Jämtland in Sweden with region 4 of Angelin (1854) and *etasje* 2 of Kjerulf (1865). However, Linnarsson (1872, pp. 38, 39) questioned the upper boundary of the formation in Sweden, which he found difficult to define. Further works by Linnarsson (1869, 1872, 1873, 1874, 1875a, b, 1876, 1878, 1879) presented the Swedish distribution of the Ceratopyge Limestone and correlations with contemporary strata elsewhere.

Brøgger (1882) revised the Norwegian Lower Silurian (=Ordovician) stratigraphy and introduced the term *Ceratopyge Limestone* as *etasje* 3aγ. Brøgger (1882, 1886, 1896) also correlated the Ceratopyge Limestone with contemporary strata in North America and Europe. Observing distinct similarities in the trilobite faunas, he coined it the Euloma–Niobe Fauna of the Ceratopyge Limestone (Brøgger 1896).

In Sweden, Moberg (1900) assigned the Ceratopyge Shale and Limestone to the new Ceratopyge Region and thought it impossible to separate the two units. Later Moberg & Segerberg (1906, p. 50) introduced the biostratigraphical Zone of *Apatokephalus serratus* for the Ceratopyge Limestone, and Wiman (1907) defined its upper boundary based on trilobites. The Lower Ordovician of Sweden was not much further revised until the work of Tjernvik (1956a), when the Zone of *Apatokephalus serratus* and its boundaries were properly defined and directly correlated with the Norwegian 3aγ, the Ceratopyge Limestone.

The accumulation of systematic and stratigraphical data made it evident that the modified and elaborated Norwegian stratigraphical scheme from Kjerulf (1865) needed a complete revision in terms of modern stratigraphical nomenclature (i.e. separate bio-, litho-, and chronostratigraphical units). The suggested schemes in the unpublished works by Fjelldal (1966) and Gjessing (1976a) drew further attention to the need for a complete

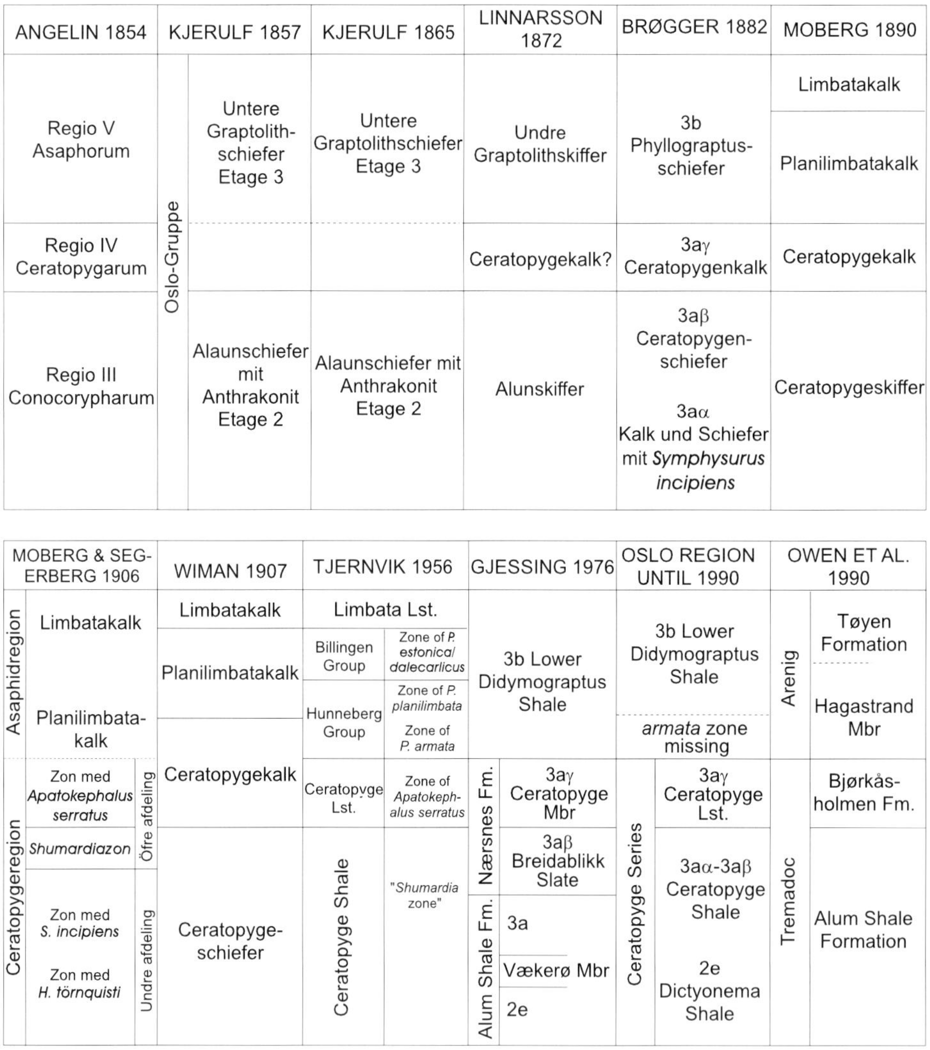

ANGELIN 1854	KJERULF 1857	KJERULF 1865	LINNARSSON 1872	BRØGGER 1882	MOBERG 1890
Regio V Asaphorum	Untere Graptolith-schiefer Etage 3	Untere Graptolithschiefer Etage 3	Undre Graptolithskiffer	3b Phyllograptus-schiefer	Limbatakalk
					Planilimbatakalk
Regio IV Ceratopygarum			Ceratopygekalk?	3aγ Ceratopygenkalk	Ceratopygekalk
Regio III Conocorypharum	Alaunschiefer mit Anthrakonit Etage 2	Alaunschiefer mit Anthrakonit Etage 2	Alunskiffer	3aβ Ceratopygen-schiefer — 3aα Kalk und Schiefer mit *Symphysurus incipiens*	Ceratopygeskiffer

(KJERULF 1857 column labelled vertically: Oslo-Gruppe)

MOBERG & SEG-ERBERG 1906		WIMAN 1907		TJERNVIK 1956		GJESSING 1976		OSLO REGION UNTIL 1990		OWEN ET AL. 1990	
Asaphidregion	Limbatakalk		Limbatakalk		Limbata Lst.		3b Lower Didymograptus Shale		3b Lower Didymograptus Shale	Arenig	Tøyen Formation
	Planilimbata-kalk		Planilimbatakalk	Billingen Group	Zone of *P. estonica/dalecarlicus*						
				Hunneberg Group	Zone of *P. planilimbata*				*armata* zone missing		Hagastrand Mbr
					Zone of *P. armata*						
Ceratopygeregion	Zon med *Apatokephalus serratus*	Øfre afdeling	Ceratopygekalk	Ceratopyge Lst.	Zone of *Apatokeph-alus serratus*	Nærsnes Fm.	3aγ Ceratopyge Mbr	Ceratopyge Series	3aγ Ceratopyge Lst.	Tremadoc	Bjørkås-holmen Fm.
	Shumardiazon			Ceratopyge Shale	"*Shumardia* zone"		3aβ Breidablikk Slate		3aα-3aβ Ceratopyge Shale		Alum Shale Formation
	Zon med *S. incipiens*	Undre afdeling	Ceratopyge-schiefer			Alum Shale Fm.	3a				
	Zon med *H. törnquisti*						Vækerø Mbr		2e Dictyonema Shale		
							2e				

Fig. 3. Historical review of the Bjørkåsholmen Formation in Scandinavia.

revision of the stratigraphy. Following the rules and recommendations for naming of geological units in Norway, issued by the Norwegian Committee on Stratigraphy (Nystuen 1986, 1989), Owen *et al.* (1990) presented a completely new stratigraphical scheme of the Ordovician in Norway. The old *etasje* system was disbanded, introducing new lithological formation names. The Ceratopyge Limestone of Norway was renamed Bjørkåsholmen Formation (Norwegian *Bjørkåsholmformasjonen*) from the type locality at Bjørkåsholmen in Slemmestad, Asker county, Norway. The formation name would also formally correspond to the concept of the Ceratopyge Limestone in Sweden, as set up from Hunneberg in Västergötland, Sweden, by Tjernvik (1956b, p. 61).

Acknowledgments. – This work is based on a Cand. Scient. thesis prepared at the Paleontological Museum, University of Oslo, Norway, in 1993. My supervisor, Professor David L. Bruton, was enthusiastic and encouraging throughout the study, offering support and constructive criticism. I also deeply acknowledge the help of the late Professor Gunnar Henningsmoen. Professor Zofia Kielan-Jaworowska (Warsaw), Dr Natascha Heintz (Oslo) and Dr. Ivar Puura (Tartu), helped with Russian and Polish literature. Fellow students and staff at Tøyen inspired and helped me during the preparation of this work, and I extend a special thank to Mr. Ole A. Hoel. During my first field season I was privileged to work in Norway with Dr. Jan Audun Rasmussen (Copenhagen). Mr. Bjørn Funke and Mr. Magne Høyberget made specimens from their private collections available. Professor John S. Peel gave me the opportunity to finish this work in Uppsala while I was supposed to concentrate on other matters. I thank Dr. Richard A. Fortey (London), for reviewing the manuscript and making a number of suggestions for its improvement. The publication is supported by a grant from the Norwegian Research Council.

I shall always be indebted to my wife Elisabeth and my mother for their support and sacrifices.

Material and methods

The main material for this study was found in the collections of the Paleontological Museum, University of Oslo. The museum houses more than three thousand samples

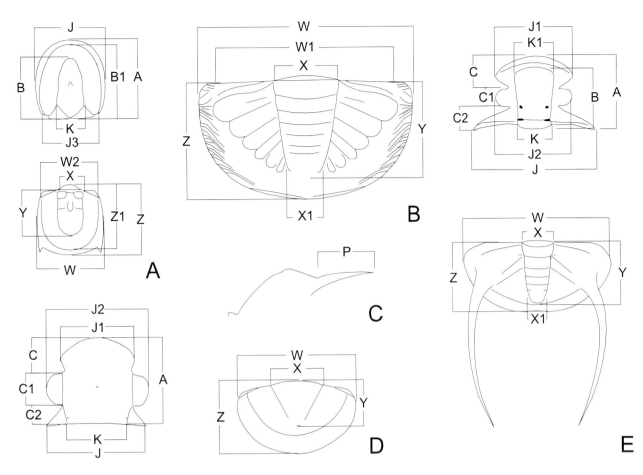

Fig. 4. Diagram showing standard measurements taken on species of representative morphologies. □A. *Geragnostus sidenbladhi.* □B. *Niobe (Niobe) insignis.* □C. *Orometopus elatifrons* profile. □D. *Nileus limbatus.* □E. *Ceratopyge acicularis.* A = length (sag.) of cranidium, B = length (sag.) of glabella, B1 = length (sag.) of cephalic acrolobe, C = length (sag.) of cranidium in front of eyes, C1 = length (sag.) of palpebral lobes, C2 = length (sag.) of cranidium behind eyes, J = posterior width (tr.) of cranidium, J1 = anterior width (tr.) of cranidium, J2 = width (tr.) of cranidium at palpebral lobes, J3 = posterior width (tr.) of cephalic acrolobe, K = posterior width (tr.) of glabella, K1 = anterior width (tr.) of glabella, P = length (sag.) of glabellar spine, X = anterior width (tr.) of pygidial rachis, X1 = posterior width (tr.) of pygidial rachis, Y = length (sag.) of pygidial rachis, Z = length (sag.) of pygidium, Z1 = length (sag.) of pygidial acrolobe, W = anterior width (tr.) of pygidium, W1 = anterior width (tr.) of pygidial pleural fields, W2 = anterior width (tr.) of pygidial acrolobe (based on Shaw 1957; Hughes 1979; Ahlberg 1989a).

from the Bjørkåsholmen Formation, collected throughout the Oslo Region during the last 140 years. To improve data on the vertical distribution and coherence of the trilobite fauna in the formation, additional material from selected localities was collected during this study.

The museum collection has remained virtually untouched for the last 30 years and has never been completely revised and registered. It is strongly biased in its contents because the bulk of the older material comes from the central Oslo area. Furthermore, only a few samples have any precise information as to the level in the unit from which they were taken. Today all samples carry a museum catalogue number and are registered in the museum database. Most of the fossils had not previously been prepared, and extensive preparation work was undertaken. Fragile specimens were strengthened by using a diluted mixture of Pioloform in alcohol. External and internal moulds yield fine details suitable for casting. Casts were prepared using Revultex latex rubber and strengthened by using a textile and fibre glass cloth backing. The rubber casts were kept in plastic boxes to avoid dust attracted by static electricity.

The smallest trilobite specimens were measured using a stereo microscope with a micrometer ocular, allowing measurements with an accuracy of 0.01 mm. Larger specimens were measured using a Vernier scale with 0.1 mm accuracy. The methods and definitions concerning the measurements follow Shaw (1957) and Hughes (1979), with a few adjustments. A standard one-way orientation of the trilobites was applied. Cranidia were placed in the presumed natural life position of each species.

The pygidia were similarly placed in the presumed natural life position of each species. Generally ten measurements were made on the cranidia and six on the pygidia (Fig. 4). Some species required additional measurements. Unfortunately, different specimens of a species cannot always be identically oriented in this manner, which gives some degree of uncertainty.

At six localities across the Oslo Region and adjacent areas, systematic sampling techniques were used to provide quantitative data on the relative vertical and regional abundance, distribution and coherence of the trilobite species throughout the Oslo Region. Older data on a seventh section were also incorporated. For each locality the UTM grid-references are given, based on topographical sheets (1:50,000, M-711 series) and economic maps (1:5000 or 1:10,000).

At each of the six localities the same procedure was followed to assure valid statistical data. The total thickness of the limestone unit was subdivided into intervals of usually 10 cm, or smaller intervals if naturally present in the unit. This assured a high sample density (S) which defines the total thickness of the individual sampled beds divided by the total thickness of the unit (Jaanusson 1976, p. 303). For some limestone horizons, 5 cm intervals were used to get a higher sample density. For each interval throughout the unit, the same type of sampling technique was applied. It was modified from the principles of the *sample frequency method* introduced by Jaanusson (1979), which is based on the presence or absence of a species in a sample represented by complete individuals or fragments. For each interval, only the first 30 (sometimes the first 50) specimens identified were included in the data set, regardless of the quantity of rock needed to get this number of specimens. This meant crushing the samples into smaller fragments to reveal as many trilobite fossils as possible.

Sample size is usually resolved after the manner of sampling and technique of inference has been selected. Jaanusson (1979, p. 253) emphasized four main effects on the accuracy of the sampled data representing the species distribution: (1) The nature of the rock; (2) variation in the vertical spacing of sampled beds (sample density); (3) variation in sample size, and (4) differences in the collecting methods. The following effects also apply for the Bjørkåsholmen Formation: (1) Thin intercalations of shale exist at several levels in the unit, but none of these was the target of investigation. The limestone itself is relatively homogenous throughout the formation. (2) The sample density is high because of the stratified sampling technique applied at the localities. In the thin limestone

unit of the Bjørkåsholmen Formation, much information would be lost if the sample density was low. (3) The sample frequency method (Jaanusson 1979) is very well suited for a technique where the rock is crushed into small bits in search of fossils. The method assumes a certain uniformity in sample size, because the frequency of samples in which a species occurs is proportional to the average frequency of the species. That is, a species recorded from a great number of samples is on average more common than a species recorded from a smaller number of samples. Sample size as referred to by Jaanusson (1979) is the actual amount of rock collected. In this study the sample size is the number of trilobites collected instead of the amount of rock sampled. This is valid, because the intention of the sampling is to make inference only about the relative species distribution in a relatively homogenous and very thin lithological unit. Furthermore, where the sample density is high, the trilobite fauna is very well known and the data set is easily reproducible from one locality to another. (4) The same sampling technique was applied for all the investigated sections.

Still, the sampling methods employed inevitably introduce a number of uncertainties, resulting in data that are less suitable for a quantitative approach to the inferred problems (see also Nielsen 1995, p. 49, for discussions of these problems). The sample-frequency method implies that each trilobite fossil or fragment revealed is counted as one, discriminating fragments derived from crushing of larger exoskeleton parts already included in the count. Furthermore, ideally all fossils within a certain volume of rock should be recorded. This is practically impossible, since disintegrating the rock means destroying some fossils, and not all fragments are easily recognizable. The actual number of specimens lost during investigation is impossible to estimate. Combine this with the size, morphology and fragility of tests, and the bias is strengthened. Small or fragile species are likely to be greatly under-represented in the recorded material. One example is the genus *Harpides*, which seems to be a thin-shelled, fragile form of which only a few fragmentary cephala are known. Usually only small fragments are recognized. The minute genus *Shumardia* is easily overlooked when the rocks are crushed and is therefore most likely highly under-represented in the counts, while *Symphysurus*, having a robust test, is more easily preserved and detected. Another strong bias may result from transportation, although for the Bjørkåsholmen Formation this seems not so significant, as discussed in a later section.

Although the faunal logs from the six sampled sections are all based on these sampling approaches, the recorded trilobite densities should be interpreted with caution. The information extracted is useful as a guide to the distribution, but does not represent its actual nature.

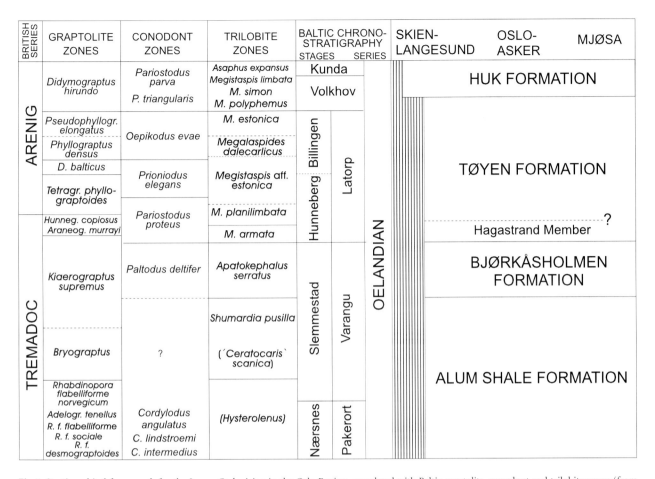

Fig. 5. Stratigraphical framework for the Lower Ordovician in the Oslo Region, correlated with Baltic graptolite, conodont and trilobite zones (from Löfgren 1993; Erdtmann & Pallits 1995; Nielsen 1995).

Stratigraphy

More than sixty localities across the Oslo Region and in the Lower Allochthon were registered in this study, and most of them were logged and measured. A detailed regional stratigraphical correlation is presented elsewhere (Ebbestad 1997), and only a short account of the setting is given here (Fig. 5).

A modern stratigraphical classification of the Ordovician strata of the Oslo Region was presented by Owen *et al.* (1990), completely revising the older *etasje* system. They introduced the name Bjørkåsholmen Formation to replace the older bio-litho stratigraphical concept of the Ceratopyge Limestone (3aγ). The older synonym is still to some extent applied to the Swedish equivalent, as set up at Hunneberg, by Tjernvik (1956b, p. 61). However, the modern formation concept also formally corresponds to the Ceratopyge Limestone in Sweden, and it should be applied throughout.

The basal part of the Bjørkåsholmen Formation is defined as the abrupt change from the dark-grey shale of the underlying Alum Shale Formation. At all the recorded localities in Norway, the transition is marked by one or more thin (5–15 cm), pale gray micritic limestone beds/nodular beds containing the Ceratopyge fauna, here understood as the fauna associated with the species *Ceratopyge forficula* or *C. acicularis* of the upper Alum Shale and Bjørkåsholmen formations, respectively. Then follows an interval of alternating light and dark shale of varying thickness (5–15 cm), with distinct dark limestone nodules (2–5 cm thick), containing a fauna dominated by the olenid *Bienvillia angelini*. This level is here considered to be of great importance for the regional correlation. The overlying beds comprise pale grey, irregularly bedded micritic or intrasparitic limestone (2–40 cm thick) with thin (1–5 cm) intercalations of shale (Bjørlykke 1974; Owen *et al.* 1990). The upper boundary of this formation is also marked by an abrupt change from mostly glauconitic limestone to the shales of the overlying Tøyen Formation. The glauconite-like sand grains occur in the Vestfossen and Oslo–Asker districts. Further north in the Oslo Region these beds disappear (Fjelldal 1966). The thickness of the formation ranges from 1.2 m at the stratotype section at Bjørkåsholmen, Slemmestad, to 0.7 m at the hypotype section at Øvre Øren, Modum (Owen *et al.* 1990).

The formation comprises the trilobite Biozone of *Apatokephalus serratus* erected by Moberg & Segerberg (1906) but properly defined by Tjernvik (1956a). The base of the graptolite Biozone of *Kiaerograptus supremus* is defined somewhere within the formation and its top somewhat higher in the overlying Tøyen Formation (Lindholm 1991). The top of the conodont Biozone of *Paltodus deltifer* coincides with the top of the Bjørkåsholmen Formation (Lindström 1971; Löfgren 1993; Erdtmann 1995; Erdtmann & Paalits 1995).

The underlying Alum Shale Formation is a widely recognizable unit across Baltoscandia. It ranges from the Middle Cambrian to the Lower Ordovician but has not yet been formally defined (Owen *et al.* 1990). Directly below the Bjørkåsholmen Formation, the Alum Shale Formation is a poorly fossiliferous unit of alternating dark- and light-grey shales, formerly known as the Ceratopyge Shale (3aβ). Gjessing (1976a, pp. 112, 114) collected a large fauna with elements of the Ceratopyge fauna from limestone horizons. A more common trilobite is *Shumardia* (*Conophrys*) *pusilla* (Sars, 1835), identifying the Biozone of *Shumardia* (*Conophrys*) *pusilla* for this unit. The trilobite fauna found in the upper part of the Alum Shale Formation has been generally accepted as equivalent to the Ceratopyge fauna, but it has been little studied. There is, however, a distinct difference on the species level between these two units.

The base of the overlying Tøyen Formation is a dark-grey shaly unit in the Oslo Region. In Sweden it is recognized in the Hunneberg area of Västergötland and in Scania. The total thickness of the formation in Norway ranges from 7.5 m in the Modum district to about 20 m in the Oslo–Asker districts (Owen *et al.* 1990). The formation is subdivided into the lower Hagastrand Member and the upper Galgeberg Member. At some localities the basal Hagastrand member forms a thick limestone unit up to 115 cm, separated from the Bjørkåsholmen Formation by a shale interval up to 30 cm. The change in the trilobite fauna is abrupt, compared to the underlying Bjørkåsholmen Formation, and allows correlation with the Swedish Biozone of *Megistaspis armata/planilimbata* of the lower Hunneberg Substage of the Latorp (Tjernvik 1956a; Erdtmann 1965b, 1995; Owen *et al.* 1990; Erdtmann & Paalits 1995).

The stratigraphical importance of the Bjørkåsholmen Formation has long been appreciated, and it has been widely discussed as reflecting an event (see Erdtmann & Paalits 1995 and references). Erdtmann (1986) defined the Ceratopyge Regressive Event (CRE) at the top of the

Bjørkåsholmen Formation. The Oslo–Scania–Lysogore confacies belt (Erdtmann 1965b; Jaanusson 1976; Erdtmann & Paalits 1995), represented in Norway by the Oslo–Asker district, includes an apparently complete section of Middle Cambrian to upper Tremadoc beds (i.e. the Bjørkåsholmen Formation in Norway). The transitional beds to the overlying strata are seen as greenish grey clay or siltstone with rhythmically intercalated thin black shale horizons (Erdtmann & Paalits 1995). In the Eiker and Modum districts this transition should accordingly be represented by the Hunneberg–Modum subconfacies belt, characterized by a distinctly reduced Tremadoc black-shale development. Here the Bjørkåsholmen Formation is instead succeeded by the condensed glauconitic *Megistaspis planilimbata* Limestone of the basal part of the Tøyen Formation. It may be noted that a similar development is found in the beach section at Vækerø Mansion, where the transitional beds of the Bjørkåsholmen and Tøyen Formation consist of glauconitic limestones with intervening shale (Owen *et al.* 1990). The thickness of the lower Tremadoc shale here is approximately 14 m (Gjessing 1976). At the hypostratotype at Øvre Øren in the Modum district, the only well-developed, unmetamorphosed section known so far there (Wöltje 1989; and personal observations 1992, 1993), faunal elements of the Zone of *Megistaspis planilimbata* were found during this study in the upper 20 cm of the unit. Glauconitic limestone seems not to be developed here at all, and the entire limestone unit was treated as the Bjørkåsholmen Formation by Owen *et al.* (1990). In the Oslo Region, development of glauconitic limestone is absent in the upper beds of the Bjørkåsholmen Formation outside the Eiker–Sandsvær and Oslo–Asker districts. Thus, in the Eiker district and partly in the Oslo district, transitional beds of the Hunneberg–Modum subconfacies type is developed. In the remainder of the Oslo–Asker district, glauconitic limestone is developed in the upper beds of the Bjørkåsholmen Formation with direct transition to overlying shale. In the Modum district no glauconitic limestone seems to be developed and the fauna of the Biozone of *M. planilimbata* is incorporated in the Bjørkåsholmen Formation. In the remainder of the Oslo Region the transition consists of limestone directly in contact with overlying shale.

For the time being the Hunneberg–Modum Subconfacies belt is used as defined by Erdtmann & Paalits (1995), but some modifications may be needed.

Faunal distribution and abundance

The wide regional distribution of the Bjørkåsholmen Formation in the Oslo Region is matched by an equally wide distribution of its trilobite fauna. The trilobite community is remarkably coherent, the same elements occurring

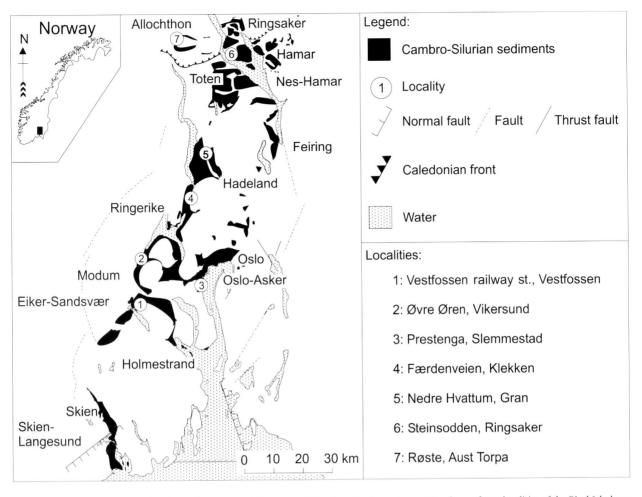

Fig. 6. Map of the Oslo Region, showing major structural elements and Cambro-Silurian outcrops. Numbers refer to localities of the Bjørkåsholmen Formation studied for relative trilobite abundance.

at most of the investigated localities. By using a combination of the Paleontological Museum (Oslo) database and data collected during extensive fieldwork, it is possible to give a general picture of the vertical and regional distribution and relative abundance of this trilobite fauna in Norway. A study of relative abundance of taxa was undertaken at six localities in different districts of the Oslo Region and adjacent districts (Fig. 6): Vestfossen railway station at Vestfossen in the Eiker–Sandsvær district (NM 4872 2217), Øvre Øren at Modum in the Modum district (NM 5763 4442), Prestenga bus stop at Slemmestad in the Oslo–Asker district (NM 8303 2724), Færdenveien at Klekken in the Ringerike district (NM 7416 7143), Nedre Hvattum at Gran in the Hadeland district (NM 8791 9382), and Røste in Aust Torpa in the Lower Allochthon Nappe district (NN 6145 4617). Data for trilobite distribution in a seventh section, Steinsodden at Moelv in the Ringsaker district (NN 9185 5368), were added using the information from Fjelldal (1966).

The Toten, Feiring and Nes–Hamar districts are not discussed here owing to scarcity of data. In the Skien–Langesund, the Bjørkåsholmen Formation is absent because of a major hiatus in that area (Henningsmoen 1960; Harper & Owen 1983; Ribland-Nilssen 1985). The seven localities give a broad and representative picture of the different development of the Bjørkåsholmen Formation in Norway (Fig. 7). At all localities the dark limestone concretions containing *Bienvillia angelini* represent an important marker horizon. However, the fauna there was not examined in detail or incorporated in the abundance study, mainly because of limited and scarce material and the evident dominance of *B. angelini*.

The faunal logs display the range of the four most common species: *Ceratopyge acicularis, Apatokephalus serratus, Symphysurus angustatus* and *Euloma ornatum*. A fifth column represents *Niobe* spp. A sixth column records the rare occurrence of agnostids, represented by *Geragnostus sidenbladhi, G. crassus* and *Arthrorhachis mobergi*. The last

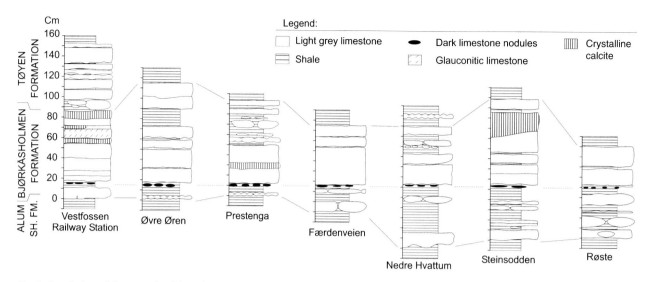

Fig. 7. Correlation of the seven localities where trilobite abundance data were collected. The level of the dark limestone nodules is taken as the correlation datum.

column contains all remaining taxa encountered. These are specified in the text. At Modum, an extra column for asaphids was added, including specimens of *Promegalaspides*, *Megistaspis* and *Niobe*.

The sample density of the formation is high. Chosen sample intervals are 10 cm, which often corresponds to the thickness of individual units. At the locality Prestenga bus stop in Slemmestad, 5 cm intervals were used, which yielded more detailed information and subtle distribution patterns. At Færdenveien at Klekken, 15 cm intervals were used for the middle unit.

Vestfossen, Eiker–Sandsvær district

Fig. 8

The fine profile at Vestfossen railroad station, Vestfossen, represents the Eiker–Sandsvær district. Here the Bjørkåsholmen Formation is 78 cm thick, with fossils present in the lower 50 cm. One basal limestone bed is present. The *Bienvillia angelini* marker bed is found in shale, 18 cm above the base, followed by four individual limestone beds with thin (1–2 cm) intercalations of shale. The two lower beds are approximately 5 cm each, and the two upper approximately 10 cm each. Above these beds follow crystalline calcite horizons and glauconitic limestone devoid of fossils.

In the basal limestone, there is about equal percentages of *Ceratopyge acicularis*, *Symphysurus angustatus*, *Euloma ornatum*, *Niobe* spp. and others (*Pliomeroides primigenus*, *Shumardia pusilla*). In the first few centimetres above the *B. angelini* bed there is a total dominance of *C. acicularis*. *Symphysurus angustatus* and *E. ornatum* are present in small numbers. In the next few centimetres *C. acicularis* is

reduced in abundance, while *A. serratus* dominates. Small numbers of *E. ornatum*, *Niobe* spp. and agnostids are present. In the two successive layers *C. acicularis* is absent and *A. serratus* is scarce, while *S. angustatus* and *E. ornatum* increase in number and become dominant. *E. ornatum* decreases somewhat in number towards the top. Other species numbers are unchanged. Forms such as *Orometopus elatifrons*, *Pagometopus gibbus*, *Shumardia pusilla* and *Pliomeroides primigenus* together constitute less than 10%.

Øvre Øren, Modum district

Fig. 9

The hypostratotype at Øvre Øren, Modum, represents the Modum district. Here the Bjørkåsholmen Formation is 118 cm thick, with fossils of the Ceratopyge fauna present in the lower 61 cm. One or maybe two nodular basal beds exist but were not investigated since they are scarce. The *B. angelini* marker horizon occurs in shale 10 cm above the base. Above the dark limestone nodules, six individual limestone beds occur. The lowermost bed is approximately 10 cm, followed by a unit nearly 20 cm thick. The two following beds are each somewhat less than 10 cm thick. Succeeding this is a 12 cm shaly unit, followed by the 34 cm thick upper limestone bed that may be divided into a lower and upper unit. This upper unit contains faunal elements of the Biozone of *Megistaspis planilimbata*, hitherto not recognized in this district. However, glauconitic limestone is not evident in the unit, and a distinction of Bjørkåsholmen Formation and the Tøyen Formation cannot be established on

Fig. 8. Trilobite abundance distribution at Vestfossen railway station (NM 4872 2217) in Vestfossen, Øvre Eiker district.

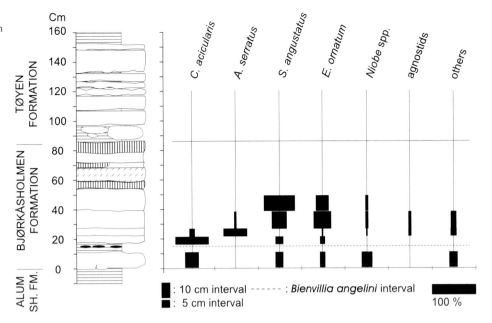

Fig. 9. Trilobite abundance distribution at Øvre Øren (NM 5763 4442) near Vikersund ski jump, Modum district.

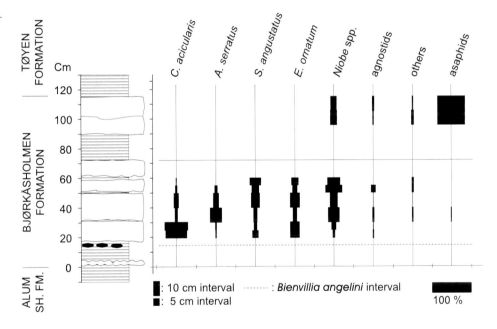

lithological grounds within the limestone sequence. The whole limestone unit is therefore referred to as the Bjørkåsholmen Formation until further investigations can be undertaken.

In the few centimetres above the unit of *Bienvillia angelini*, there is a total dominance of *Ceratopyge acicularis* with minor elements of *Apatokephalus serratus*, *Symphysurus angustatus*, *Euloma ornatum*, *Niobe* spp. and agnostids. In the following unit, *C. acicularis* is scarce while *A. serratus* dominates. Other common species remain the same: *Pagometopus gibbus*, *B. angelini* and *Promegalaspides intactus*. In the upper fossiliferous units, *C. acicularis*

and *A. serratus* are reduced in numbers, while *S. angustatus*, *E. ornatum* and *Niobe* spp. increase. The latter species dominates in the top layers. Agnostids and others (*B. angelini*, *Pagometopus gibbus* and *Harpides rugosus*) are present in small numbers.

The basal 15–20 cm of the uppermost limestone unit yielded no fossils, while the succeeding 10–15 cm is totally dominated by species of the Biozone of *Megistaspis planilimbata* including *M.* (*Paramegistaspis*) *planilimbata*, *M.* (*Lannacus*) *nericiensis*, *Promegalaspides* (*Borogothus*) *stenorachis*, *Niobe* (*Niobella*) *bohlini* and others; the Biozone of *M. armata* has not been recognized (Hoel 1999).

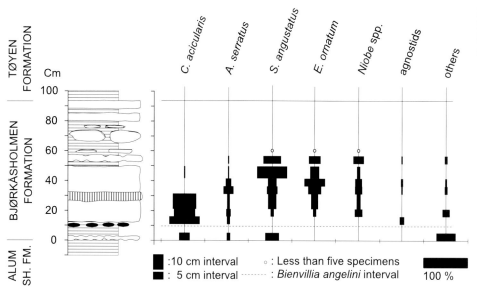

Fig. 10. Trilobite abundance distribution at Prestenga (NM 8303 2724) in Slemmestad, Oslo–Asker district.

Slemmestad, Oslo–Asker district

Fig. 10

The outcrop at Prestenga bus stop, Slemmestad, represents the Oslo–Asker district. Here the Bjørkåsholmen Formation is 95 cm thick with fossils present in the lower 58 cm. One thin (5 cm) nodular limestone bed is developed at this locality. Some 100 m to the north another section shows a 15 cm thick basal bed with incorporated septarian nodules. The *Bienvillia angelini* marker bed at Prestenga is present in shale 9 cm above the base. The succeeding limestone bed is approximately 30 cm thick, with a 10 cm thick crystalalline calcite horizon near the top. Above this thick bed follows a nearly 10 cm thick limestone bed, succeeded by 5 cm glauconitic limestone. This is followed by some 10 cm of shale, a thin nodular limestone, and a 10 cm thick limestone bed capping the unit.

In the basal bed, *Ceratopyge acicularis*, *Symphysurus angustatus* and *Shumardia pusilla* dominate and are present in about equal numbers. *Apatokephalus serratus*, *Pliomeroides primigenus* and *B. angelini* are present in small numbers. In the few centimetres above the layer of *B. angelini*, there is a total dominance of *C. acicularis* with a few specimens of the other more common trilobites. *P. primigenus*, *Orometopus elatifrons* and *B. angelini* represent rare species. In the following unit, *C. acicularis* is absent while *A. serratus*, *S. angustatus* and *Euloma ornatum* increase in number. Agnostids and other trilobites (*Pagometopus gibbus*, *O. elatifrons*) are scarce. In the upper fossiliferous beds, *A. serratus* decreases in number and disappears while *S. angustatus* and *E. ornatum* increase in number and become dominant. The latter species decreases in number towards the top. *Niobe* spp. are present throughout the unit in small numbers. Agnostids and others (*Harpides rugosus* and *Promegalaspides intactus*) are scarce.

Klekken, Ringerike district

Fig. 11

The locality at Færdenveien, Klekken, represents the Ringerike district. Here the Bjørkåsholmen Formation is 94 cm thick with fossils in the lower 80 cm. Two nodular limestone beds, separated by somewhat less than 10 cm of shale, make up the base. The lower bed is approximately 10 cm thick and locally developed. The second bed is approximately 5 cm thick. The *Bienvillia angelini* bed is present in shale 29 cm above the base. Above the dark limestone nodules follows an approximately 35 cm thick compact limestone bed. Succeeding this are two 10 cm thick limestone beds with 1–2 cm shale partings.

Ceratopyge acicularis and *Apatokephalus serratus* dominate the basal bed, with a small number of *Niobe* spp., agnostids and others, including *Orometopus elatifrons*. In the beds above the dark limestone nodules, *C. acicularis* dominates with small numbers of *A. serratus*, *Symphysurus angustatus*, *Euloma ornatum* and *Niobe* spp. Other species are represented by *O. elatifrons*. The sample density is low in the lower and middle part of the formation. Here *C. acicularis* decreases steadily in number. *A. serratus* is dominant. *S. angustatus*, *E. ornatum* and *Niobe* spp. are dominant in the upper beds of the formation. Agnostids, *Falanaspis aliena*, *Promegalaspides intactus* and *Pagometopus gibbus* exist in small numbers.

Jaren, Hadeland district

Fig. 12

The only available locality at Nedre Hvattum, Gran, represents the Hadeland district. Here the Bjørkåsholmen Formation is 115 cm thick, with fossils in the lower 103 cm. Two basal limestone beds are present,

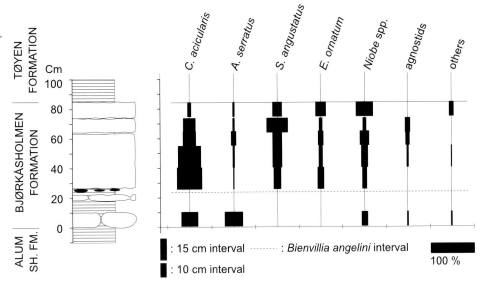

Fig. 11. Trilobite abundance distribution at Færdenveien (NM 7416 7143) in Klekken, Ringerike district.

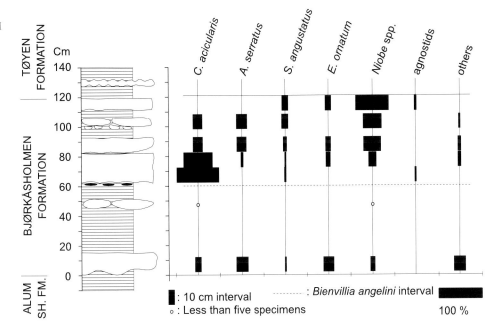

Fig. 12. Trilobite abundance distribution at Nedre Hvattum (NM 8791 9382) in Gran, Hadeland district.

the lower one 15 cm thick, and the upper nodular and somewhat more than 5 cm thick. Approximately 30 cm of shale separates them. The *Bienvillia angelini* marker horizon is present in the shale at 55 cm. Directly above follows first a 20 cm thick bed overlain by another 10 cm thick bed. The succeeding two uppermost beds are nodular, each somewhat less than 10 cm thick and parted by 5 cm of shale.

The basal unit has about equal numbers of *Ceratopyge acicularis, Apatokephalus serratus, Euloma ornatum, Niobe* spp. and others (*B. angelini*). *Symphysurus angustatus* is

scarce. In a thin nodular limestone bed 30 cm above the base, *C. acicularis* and *Niobe* spp. were found. In the beds above the dark limestone nodules, *C. acicularis* dominates with few specimens of other species. In the following bed, *C. acicularis* decreases in number but is still common together with *A. serratus* and *Niobe* spp. The latter species increases steadily in number and dominates totally towards the top of the formation. *S. angustatus* and *E. ornatum* increase only slightly in numbers towards the top. Agnostids and others (*Orometopus elatifrons* and *B. angelini*) are scarce.

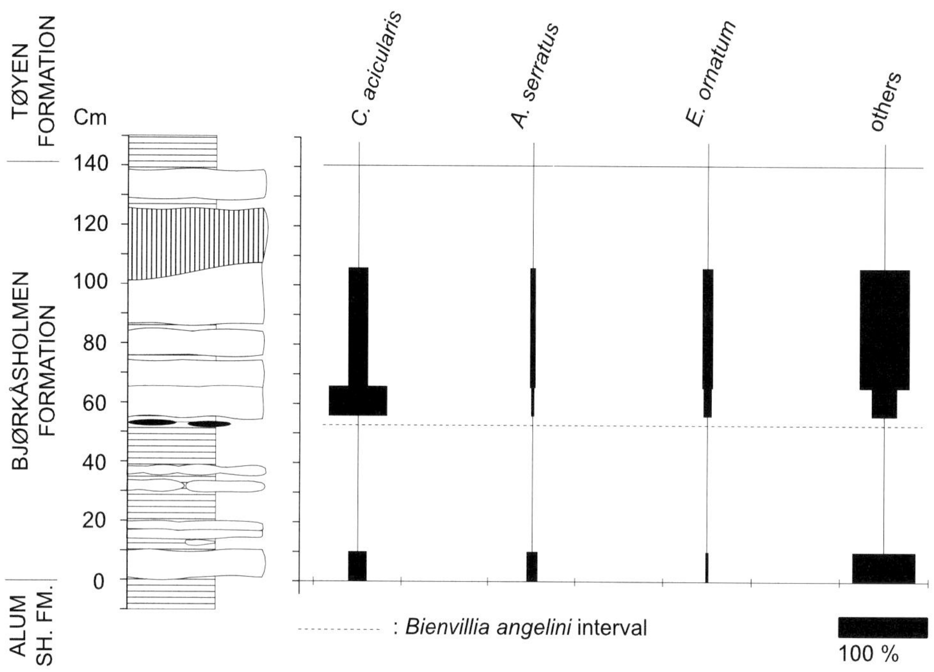

Fig. 13. Trilobite abundance distribution at Steinsodden (NN 9185 5368) in Ringsaker, Ringsaker district (based on unpublished data from Fjelldal 1966).

Steinsodden, Ringsaker district

Fig. 13

The locality at Steinsodden, Ringsaker, represents the Ringsaker district, based on the data of Fjelldal (1966). The locality is now protected and cannot be sampled.

The Bjørkåsholmen Formation is here 140 cm thick, with fossils throughout. The basal beds are developed as several thin (5–10 cm) limestone horizons, parted by shale of varying thickness. The *Bienvillia angelini* marker bed is present 55 cm above the base of the formation. Directly above this bed follows three 10 cm thick limestone beds succeeded by an approximately 35 cm thick limestone bed, with crystalline calcite developed in the upper 20–25 cm. An uppermost 10 cm thick limestone bed ends the unit.

In the basal beds, *Ceratopyge acicularis* and *Shumardia pusilla* dominate, with small numbers of *Apatokephalus serratus*, *Symphysurus angustatus*, *Euloma ornatum* and *Niobe* spp. *Pliomeroides primigenus* and *B. angelini* are also present in small numbers. In the unit above the dark limestone nodules, *C. acicularis* dominates, with small numbers of the other common species. Other species are represented by agnostids, *Shumardia pusilla* and *Orometopus elatifrons*. The sample density in the following units is low, but it is obvious that *C. acicularis* decreases markedly in numbers while the other common species relatively increase. Other species are represented by agnostids and *Promegalaspides*.

Torpa, Lower Allochthon

Fig. 14

The locality at Røste, Aust Torpa, represents the Lower Allochthon Nappe district. Here the Bjørkåsholmen Formation is 91 cm thick, with fossils in the lower 80 cm. Three limestone beds of approximately 10 cm thickness each form the base of the unit. They are partly nodular. The *Bienvillia angelini* marker horizon is present 48 cm above the base. The upper part of the unit consists of two limestone beds, each approximately 15 cm thick with a 5 cm shale parting.

The data are somewhat limited for the lower beds, but *Ceratopyge acicularis*, *Apatokephalus serratus*, *Symphysurus angustatus*, *Niobe* spp., *Pliomeroides primigenus* and *B. angelini* are represented. *C. acicularis* dominates in the unit above the dark limestone nodules with a few specimens of *A. serratus*, *Euloma ornatum*, *P. primigenus* and *Dikelokephalina dicraeura*. The data are also limited in the upper unit; no specimens of *C. acicularis* were found while *A. serratus* and *Niobe* spp. dominate. *P. primigenus* is also very common.

Summary of faunal logs

The relative abundance above the dark limestone nodules with *Bienvillia angelini* is fairly consistent in all the sections investigated. In the basal limestone beds, which are more numerous northwards in the Oslo Region, a relative abundance cannot be established.

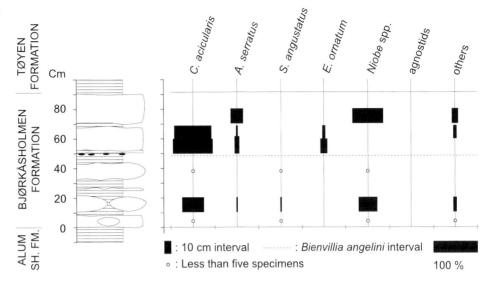

Fig. 14. Trilobite abundance distribution at Røste (NN 6145 4617) in Aust Torpa, Lower Allochthon district.

As stated earlier, the occurrence of the black limestone nodules containing *B. angelini* is an important stratigraphical event. The shale in which these nodules are embedded is sometimes black, indicating high contents of organic matter and originally anaerobic conditions. Systematic revision has also shown that the trilobite fauna encountered in these nodules is mostly specifically different from that in the remainder of the unit. The trilobite fauna is completely dominated by *B. angelini*, with minor elements of *Niobe* (*Niobella*) *obsoleta*, *Shumardia* (*Conophrys*) *pusilla*, *Ceratopyge* sp., *C. acicularis* and *Nileus limbatus*. The ostracod *Nanopsis nanella*, the brachiopods *Archaeorthis christiania*, *Broeggeria salteri* and *Acrotreta?* sp., the gastropod *Sinuitella norvegica*, and rare specimens of the cephalopod *Orthoceras attavus* are also encountered. These species are usually confined to this level. The quite diverse fauna and the facies regime therefore represents an anomaly in the predominantly light-shale/light-limestone facies at this time. The event is also recorded in all the districts investigated but may not be synchronous.

The upper stable limestone facies shows a relatively coherent abundance distribution throughout. In the light grey limestone bed succeeding the *Bienvillia*-horizon, there is a dominance of *C. acicularis*. Upwards, *C. acicularis* decreases in number and is replaced by a relative increase in the number of *A. serratus* specimens together with or slightly before the acme of *E. ornatum*. This is most evident at the localities Vestfossen and Modum. In Slemmestad this situation is present to a subtle degree. *Symphysurus angustatus* shows a steady increase in number, with an acme succeeding that of *E. ornatum*, then followed by a relative decrease in number, as shown both at Slemmestad and at Klekken. *Niobe* spp. are prominent components throughout, with a slight increase in number at all localities except at Steinsodden, where the data are poor.

Notes on biostratigraphy and palaeogeography

The fossil remains of the Bjørkåsholmen Formation are dominated by trilobite exuviae. The fragments are randomly oriented in the sediment and unsorted. Cavities beneath curved tests have not been found. Disarticulated exuviae dominates, but some relatively complete trilobite moults are known. Local accumulations of unsorted, mostly unfragmented, trilobite exuviae are also common. Speyer (1987, p. 211) found that this mode of preservation usually indicates deep intrastratal bioturbation. It therefore seems that bioturbation was the main agent in homogenizing the sediment of the Bjørkåsholmen Formation. Presumably only limited transportation by currents or waves occurred with a mostly autochthonous formation of the sediment. The presence of a few, almost complete articulated specimens points to a short surface exposure of the dead animals and a usually limited destruction by bioturbation.

A basic facies pattern of a carbonate unit like the Bjørkåsholmen Formation is that of a shallow and tectonic stable shelf environment (Wilson 1975, pp. 25, 355–356). The water depth would reach below normal wave base and even below storm wave base. The sea water would be oxygenated, with normal salinity and good current circulation. The sediments would generally be uniform and widespread, well-homogenized, very fossiliferous with diverse shelly fauna, and have wavy and nodular beds and load cast structures. These features match well with the setting of the Bjørkåsholmen Formation, at least for the post-*Bienvillia* beds. For the basal beds, some factors suggest a higher degree of mixing and perhaps resedimentation. In the Oslo–Asker district septarian nodules are seen in the basal bed/beds at several localities, containing the same fauna as the surrounding matrix. The precence of these structures indicates an early period of lime-mud development and lithification, followed by resedimentation of the septarian nodules.

Further north in the Oslo Region, the number of basal beds increases without giving better resolution in terms of trilobite abundance distribution. The sequence of altering limestone beds and shale below the dark nodules may indicate repeated initiation of the Bjørkåsholmen Formation facies in a period of unstable basin settings, variations in depth or pycnocline, before the establishment of a stable deposition environment represented by the *B. angelini* bed and the succeeding limestone facies. The trilobite abundance also reflects the stable sedimentary environment throughout the upper part of the unit. The composition of the limestone beds varies, even on a very local scale, with calcite horizons, abrupt changes between limestone and the intercalating shales, glauconitic limestone and varying thickness. At all localities the number of trilobites decreases steadily upwards in the upper part of the unit, because the amount of rock needed to extract the same number of trilobites throughout increased upwards for all the sections.

The trilobite fauna of the Bjørkåsholmen Formation comprises more than thirty species. The most common ones, *Ceratopyge acicularis*, *Symphysurus angustatus*, *Euloma ornatum*, *Apatokephalus serratus* and *Niobe* spp., in order of decreasing abundance, are found in all districts in Southern Norway. The other species are rare and not represented in every district. Several of the common species are represented by complete growth series, and also within the uncommon taxa distinct differences in size may be seen. This shows a picture of a widespread, diverse and stable trilobite fauna. Brachiopods are another common fossil group in the formation, and the characteristic microbrachiopod assemblage consists of a mixture of relict Cambrian taxa as well as the earliest representatives of Ordovician lingulids and acrotretids (Popov & Holmer 1994, p. 31).

Following the CRE (Erdtmann 1986), all except three species of the Bjørkåsholmen Formation trilobite fauna disappeared. The surviving species were *S. angustatus*, *Falanaspis aliena* and *Promegalaspides stenorachis*. The two latter species are not very prominent components of the Ceratopyge fauna, and *S. angustatus* seems to have been a very tolerant species, though it is not very abundant after the CRE. On the generic level, *Geragnostus*, *Arthrorhachis*, *Shumardia*, *Euloma*, *Saltaspis*, *Apatokephalus*, *Niobe* (*Niobe*), *Niobe* (*Niobella*), *Promegalaspides*, *Nileus*, *Symphysurus*, *Varvia*, *Ottenbyaspis*, *Orometopus*, *Falanaspis* and *Agerina* continued into the succeeding strata in Scandinavia. Thus, it is correct to say that a complete species turnover occurred at the CRE with an extinction of the Ceratopyge fauna in Baltica.

The early Tremadoc Olenid province comprised Baltica, Siberia and Gondwanaland territories and was replaced by the Ceratopyge province and Ceratopyge

fauna in the late Tremadoc (Henningsmoen 1957; Whittington & Hughes 1974). This was also a period with major diversification and origination of new trilobite genera in shallow water on open shelf environments, whereas older Cambrian taxa continued in more offshore to marginal sites (Fortey & Owens 1991). Restricted to the Ceratopyge Province were older Olenid fauna type elements with few Tremadoc species, like *Tropidopyge*, *Peltocare* and *Saltaspis*, all with a Baltic origin. Other old genera belonging to this older fauna, like *Parabolinella* and *Bienvillia*, together with the Ceratopyge fauna like *Euloma*, *Niobe*, *Shumardia* and the agnostid *Geragnostus*, had a cosmopolitan distribution. Within the newer Ceratopyge fauna type elements, the genera *Promegalaspides* (*Borogothus*), *Ceratopyge* and *Orometopus* were also restricted to the Ceratopyge Province with development in the early and late Tremadoc. Genera like *Apatokephalus*, *Agerina*, *Harpides*, *Parapilekia*, *Pliomeroides*, *Dikelokephalina*, *Symphysurus* and *Nileus* were cosmopolitan. The first five genera have first occurrences in Gondwanaland (S. America). *Dikelokephalina* spread in the late Tremadoc in Gondwanaland, Avalonia and Baltica. The genera *Symphysurus* and *Nileus* spread widely in early and late Tremadoc, respectively, along the marginal areas of platforms. The latter genus has its first occurrence in Baltica and is commonly related to a muddy substrate (Nielsen 1995). The genera *Varvia* and *Pagometopus* seem to have been endemic to Scandinavia. Small genera like *Ottenbyaspis* and *Falanaspis* are mostly confined to the Scandinavian strata. Only one report from Bohemia (Růžička, 1926) and one report from Svalbard (Fortey 1975), respectively, are known.

Thus, the Bjørkåsholmen Formation trilobite fauna comprises a few endemic genera, some Ceratopyge fauna elements with essentially Baltic development, and, finally, mostly cosmopolitan genera with probable South American origin.

Fig. 15 presents the stratigraphical distribution in Norway and Sweden of the species discussed herein.

Fig. 15. Stratigraphical distribution of the species discussed herein, including data from Norwegian and Swedish strata. Based on Tjernvik (1956a), Ahlberg (1992), Nielsen (1995), Hoel (1999 and in press) and own data.

Species \ Biozones	Zone of Shumardia pusilla	Level of Bienvillia angelini	Zone of Apatokephalus serratus	Zone of Megistaspis armata	Zone of Megistaspis planilimbata
Geragnostus sidenbladhi (Linnarsson, 1869)	■■■		■■■		
Geragnostus crassus Tjernvik, 1956		■	■■■		
Arthrorhachis mobergi (Tjernvik, 1956)		■	■■■		
Euloma ornatum Angelin, 1854	■■■■				
Hospes? sp.			■■		
Shumardia sp.					
Shumardia (Conophrys) pusilla (Sars, 1835)	■■■■■				
Hypermecaspis rugosa Brøgger, 1882		■ ?			
Tropidopyge broeggeri (Moberg & Segerberg, 1906)		?	■■		
Peltocare modestum Henningsmoen, 1959			■■		
Saltaspis stenolimbatus n. sp.		■			
Bienvillia angelini (Linnarsson, 1869)		■■■			
Parabolinella lata Henningsmoen, 1959			■■		
Apatokephalus serratus (Boeck, 1838)	■■■		■■		
Apatokephalus dubius (Linnarsson, 1869)			■■		
Apatokephalus dactylotypos n. sp.		■			
Apatokephalus cf. *sarculum* Fortey & Owens, 1990	■■■■				
Niobe (Niobe) insignis Linnarsson, 1869		■ ?	■■		
Niobe (Niobella) obsoleta Linnarsson, 1869		? ?	■■		
Niobe (Niobella) eudelopleura n. sp.		■			
Promegalspides (Borogothus) intactus (Moberg & Segerberg, 1906)	■■■■■		■■		
Promegalaspides (Borogothus) stenorachis (Angelin, 1851)			■■■■■■■		
Ceratopyge forficula (Sars, 1835)	■■■	?			
Ceratopyge acicularis (Sars & Boeck, 1838)			■■		
Ceratopyge sp.	■■■■				
Dikelokephalina dicraeura (Angelin, 1854)			■■		
Nileus limbatus Brøgger, 1882			■■■■■■		
Symphysurus angustatus (Boeck, 1838)		■	■■■■■■		
Varvia longicauda Tjernvik, 1956			■■■		
Ottenbyaspis oriens (Moberg & Segerberg, 1906)			■■		
Orometopus elatifrons (Angelin, 1854)		■			
Pagometopus gibbus Henningsmoen, 1959			■■		
Falanaspis aliena Tjernvik, 1956			■■■■■■		
Agerina praematura Tjernvik, 1956			■■■		
Harpides rugosa (Boeck, 1838)			■■		
Parapilekia speciosa (Dalman, 1827)			■■		
Pliomeroides primigenus (Angelin, 1854)		■			

Systematic palaeontology

The terminology used here is that of Harrington *et al. in* Moore (1959, pp. O117–O126), with the following exceptions: The terms *rachis* and *dorsal furrow* are used instead of 'axis' and 'axial furrow', following Jaanusson (1956, pp. 36–36). The glabella is taken to include the occipital ring and the term *bacculae* is used for the areas abutting the posterior part of the glabella, following Fortey (1975, pp. 14–15). Number of pygidial rachial rings does not include the terminal piece of the pygidial rachis.

For the agnostids the terminology used by Ahlberg (1989a, p. 543 and references) is followed. Trilobite taxa of the suborder Asaphina are arranged according to the scheme used by Fortey & Chatterton (1988). Lengths and widths of specimens refer to sagittal (sag.) and transverse (tr.) directions, respectively, and are given in centimetres in the tables. Unless otherwise stated, the specimen numbers in the tables refer to samples from the collections of the Paleontological Museum in Oslo, Norway.

Occurrences of specimens in the dark limestone nodules at the base of the unit are usually pointed out specifically; if not otherwise stated the specimens occur in the light-grey limestone of the formation.

The type material for several of the species described herein could not be located. Neotypes would have been relatively easy to select for most of them. However, this was not done because of the relatively great likelihood of eventually recovering lost specimens in the museum collections. During this study some lost type specimens were actually recovered at the Paleontological Museum in Oslo. The collections at the Natural History Museum in Stockholm, Lund University and Palaeontological Museum, and the Swedish Geological Survey in Uppsala house many of the old collections associated with the Bjørkåsholmen Formation. Investigations of the collections there were not within the limits of this study, and therefore missing originals were not looked for.

The following abbreviations denote institutions where particular specimens are housed.

LO   Department of Historical Geology and Palaeontology, Lund University, Sweden.
PMO   Paleontological Museum, University of Oslo, Norway.
PMU   Museum of Evolution, Palaeontology Section, Uppsala University, Sweden.
RM   Swedish Museum of Natural History, Stockholm.
SGU   Geological Survey of Sweden, Uppsala.
NMW   National Museum of Wales, Cardiff.

Family Metagnostidae Jaekel, 1909

Genus *Geragnostus* Howell, 1935

Type species. – *Agnostus sidenbladhi* Linnarsson, 1869, pp. 82–83, Pl. 2:60, 61, from the Biozone of *Apatokephalus serratus*, Bjørkåsholmen Formation (upper Tremadoc), at Mossebo in Hunneberg, Västergötland, Sweden; by original designation of Howell (1935, p. 231).

Discussion. – The genus has recently been discussed and redefined by Fortey (1980) and Ahlberg (1989b), and discussed in the classification of Agnostida by Shergold *et al.* (1990). The importance of the length of the pygidial rachis and the position of the transglabellar furrow in distinguishing *Geragnostus* and *Arthrorhachis* were emphasized. Ahlberg (1989b, 1992) also included in the genus *Geragnostus* non-scrobiculated metagnostids with a long pygidial rachis, in which the sagittal length of the terminal lobe is longer or subequal to the combined sagittal length of the anterior and middle lobes. Intraspecific variants were also noted by the above-mentioned authors.

Geragnostus sidenbladhi (Linnarsson, 1869)

Fig. 16

Synonymy. – □1869 *Agnostus sidenbladhi* n.sp. – Linnarsson, pp. 82–83, Pl. 2:60, 61. □1906 *Agnostus sidenbladhi* Linnarsson – Moberg & Segerberg, p. 77, Pl. 4:1. □*non* 1906 *Agnostus sidenbladhi* Linnarsson – Lake, p. 22, Pl. 2:17. □1956a *Geragnostus sidenbladhi* (Linnarsson) – Tjernvik, p. 188, Text-fig. 27A, Pl. 1:5, 6. □1959 *Geragnostus sidenbladhi* (Linnarsson) – Harrington *et al. in* Moore, p. O176, *non* Fig. 115:3. □*non* 1961 *Geragnostus sidenbladhi* (Linnarsson) – Balashova, p. 112, Pl. 1:3, 4. □1980 *Geragnostus sidenbladhi* (Linnarsson) – Fortey, pp. 21, 25, 27. □1989b *Geragnostus sidenbladhi* (Linnarsson) – Ahlberg, pp. 310–316, Figs. 1A–K, 2A–C, 3A–G. □1992 *Geragnostus sidenbladhi* (Linnarsson) – Ahlberg, p. 545, Fig. 6a–f.

Lectotype. – A nearly complete pygidium (SGU 25) from the Bjørkåsholmen Formation at Mossebo in Hunneberg, Västergötland, Sweden. Selected and figured by Ahlberg (1989b, p. 310, Fig. 3A). Illustrated by Linnarsson (1869, Pl. 2:6) and Moberg & Segerberg (1906, Pl. 4:1).

Norwegian material. – Three cephala, ten pygidia.

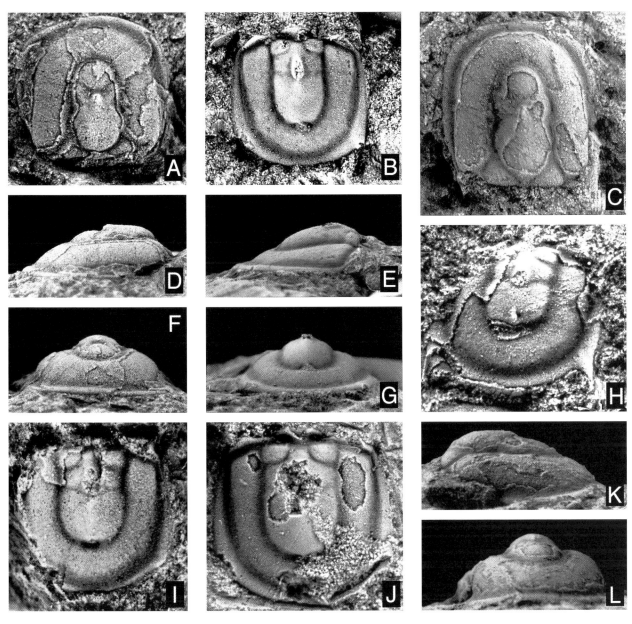

Fig. 16. Geragnostus sidenbladhi (Linnarsson, 1869). □A, D, F. Dorsal, lateral, and anterior view of cephalon with exoskeleton partly preserved; ×11, PMO 136.019. Found 35–46 cm above the base of the formation at Færdenveien in Klekken, Ringerike. Coll.: J.O.R. Ebbestad, 1992-09-06. □B, E, G. Dorsal, lateral and posterior view of pygidium (internal mould); ×11, PMO 99186/1. Bjørkåsholmen, Asker. Coll.: G. Henningsmoen, 1959-02-22. □C, K, L. Dorsal, lateral and anterior view of cephalon showing the prominent basal lobes; ×11, PMO 99185. Bjørkåsholmen, Asker. Coll.: G. Henningsmoen, 1959-02-22. □H. Latex replica from external mould of pygidium with preserved border spines; ×22, PMO 35983. Storhamar strand, Hamar. Coll.: T. Strand, 1926-08-21. □I. Internal mould of pygidium showing short posterior lobe; ×13, PMO 136.018. Found 63–73 cm above base of the formation at Færdenveien in Klekken, Ringerike. Coll.: J.O.R. Ebbestad, 1992-09-06. □J. Dorsal view of pygidium showing the transverse anterior border; ×14, PMO 121.637. 32–40 cm above the base of the formation, Øvre Øren, Modum. Coll.: J.O.R. Ebbestad, 1992-08-26.

Discussion. – This species was recently redescribed and discussed by Ahlberg (1989b, pp. 310–316, Figs. 14A–K, 15A–C, 16A–G), and a redescription is not necessary here. For discussion on comparable species, see also Ahlberg (1992). As revised, *G. sidenbladhi* is a fairly variable species, but the Norwegian specimens correspond closely to the figured Swedish material. The well-preserved Norwe-gian specimens reflect some of the intraspecific variation. The length/width ratio of the cephala (Fig. 19A) varies from 1 to 1.2 times that of the Swedish figured specimens and, as stated by Ahlberg (1989b, p. 315), the terminal lobes of the pygidial rachis are 1.3–1.6 times longer than the anterior and middle lobes combined (Fig. 16B: Long terminal lobe; Fig. 16I: Short terminal lobe). It may also be noted that the basal lobes appear to be larger, more prominent in some Norwegian specimens (Fig. 16C, K). Tables 1 and 2 give the measurements of the cephala and pygidia, respectively.

Geragnostus crassus Tjernvik, 1956

Fig. 17

Synonymy. – □1905 *Agnostus glabratus* Angelin var. *ingricus* Schmidt – Wiman, p. 13, Pl. 1:23, 24. □1906 *Agnostus sidenbladhi* Linnarsson var. *ureceolatus* Segerberg mscr. – Moberg & Segerberg, pp. 77–78, Pl. 4:4. □1956a *Geragnostus crassus* – n.sp., Tjernvik, p. 190, Text-fig. 27B, Pl. 1:7, 8. □1992 *Geragnostus crassus* Tjernvik – Ahlberg, pp. 547–548, Fig. 7a–r.

Holotype. – An internal mould of a complete pygidium (PMU Vg279) from the Bjørkåsholmen Formation at Stenbrottet in Västergötland, Sweden. Identified and illustrated by Tjernvik (1956a, p. 190, Pl. 1:8).

Norwegian material. – Eight cephala, six pygidia.

Discussion. – Ahlberg (1992, pp. 547–548, Fig. 7a–r) redescribed and discussed this species, and no redescription is necessary here. The glabellar length (sag.) is longer, nearly 80% of total cephalic length, in the Norwegian material. The convexity of the cephala and pygidia varies markedly within both the Swedish and Norwegian specimens (Fig. 17B, G). This is also reflected in the variation in the length/width ratios of the cephala, ranging from 1 to 1.2 (Fig. 17A: long cephalon; Fig. 17E: short cephalon, Fig. 19A). Tables 3 and 4 give the measurements of the cephala and pygidia, respectively.

Table 1. Cephalic measurements of *Geragnostus sidenbladhi*.

Specimen	A	B	B1	J	J3	K
S1348/2	0.23	0.13	0.21	0.19	0.20	0.08
83736/6	0.26	0.18	0.23	0.29	0.25	0.12
99185	0.28	0.20	0.26	0.25	0.24	0.10
136.019	0.31	0.21	0.31	0.31	0.28	0.11

Table 2. Pygidial measurements of *Geragnostus sidenbladhi*.

Specimen	X	Y	Z	Z1	W	W2
S1267/1	0.04	0.06	0.10	0.09	0.10	0.09
S1341/4	0.07	0.12	0.20	0.18	0.19	0.17
35983	0.07	0.10	0.18	0.15	–	0.17
97156/2	0.08	0.11	0.16	0.14	0.21	0.18
97272/1	0.06	0.12	0.17	0.16	0.14	–
99186/1	0.13	0.22	0.30	0.28	0.30	0.27
121.621/1	0.08	0.10	0.20	0.17	0.19	0.17
121.637	0.13	0.22	0.28	0.29	0.28	0.29
136.018	0.12	0.19	0.30	0.26	0.30	0.26

Table 3. Cephalic measurements of *Geragnostus crassus*.

Specimen	A	B	B1	J	J3	K
1298/2	0.30	0.21	0.27	0.24	–	0.11
1398/1	0.28	0.18	0.24	0.24	–	0.10
40591/2	0.27	0.19	0.24	0.23	0.24	0.11
64120/3	0.42	0.30	0.38	0.41	0.37	0.12
83785/1	0.25	0.16	0.22	0.12	0.12	0.10
83940/1	0.22	0.15	0.19	0.24	0.24	0.09
83940/2	0.31	0.22	0.25	0.28	0.28	0.11
136.021	0.14	0.11	0.15	0.14	0.14	0.06

Table 4. Pygidial measurements of *Geragnostus crassus*.

Specimen	X	Y	Z	Z1	W	W2
S1137/2	0.08	0.13	0.16	0.15	0.18	0.19
1102/4	0.10	0.18	–	0.24	0.27	0.26
1356/4	0.06	0.07	0.12	0.10	0.12	0.14
61480/10	0.27	0.39	0.58	0.44	0.48	0.60
61480/11	0.06	0.10	0.12	0.12	0.14	0.16
121.610/2	0.11	0.16	0.19	0.20	0.22	0.22

Genus *Arthrorhachis* Hawle & Corda, 1847

Type species. – *Battus tardus* Barrande, 1846, p. 35, from the Králův Dvůr Formation (Ashgill) at Libomyšl near Zdice, Czech Republic; by monotypy of Hawle & Corda (1847, p. 115).

Discussion. – The type species was refigured by Pek (1977, Pl. 8:2). Fortey (1980) revived *Arthrorhachis*, previously considered a junior synonym of *Trinodus* M'Coy, 1846. Ahlberg (1989b, 1992) also included in the genus *Arthrorhachis* non-scrobiculate metagnostids with a short pygidial rachis, in which the sagittal length of the terminal lobe is shorter than the combined sagittal length of the anterior and middle lobes. Shergold *et al.* (1990) presented a classification and distribution of Agnostida, were *Arthrorhachis* was discussed.

Arthrorhachis mobergi (Tjernvik, 1956)

Fig. 18

Synonymy. – □1906 *Agnostus trinodus* Salter var. – Moberg & Segerberg, p. 78, Pl. 4:5a, b. □1956a *Trinodus mobergi* n.sp. – Tjernvik, p. 195, Text-fig. 28A, Pl. 1:18, 19. □1980 *Arthrorhachis mobergi* (Tjernvik) – Fortey, pp. 27, 32. □1992 *Arthrorhachis mobergi* (Tjernvik) – Ahlberg, pp. 560–561, Fig. 16a–i.

Holotype. – A complete cephalon (PMU Vg281) from the Bjørkåsholmen Formation at Stenbrottet in Västergötland, Sweden. Identified and illustrated by Tjernvik (1956a, p. 195, Pl. 1:18).

Norwegian material. – Six cephala, two pygidia.

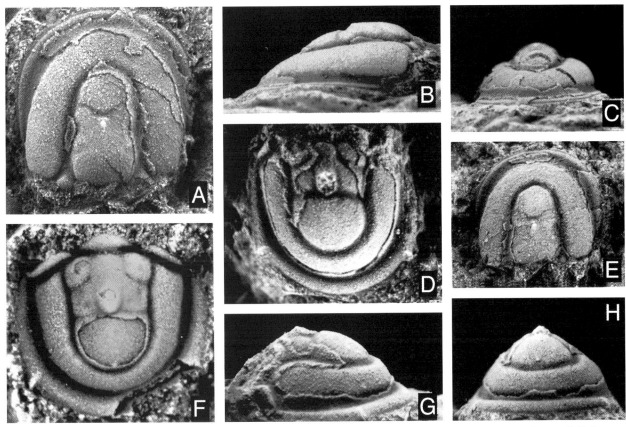

Fig. 17. Geragnostus crassus Tjernvik, 1956. □A–C. Dorsal, lateral and anterior view of internal mould of cephalon showing outlines, cephalic furrows, and a gentle convexity; A ×8, B–C ×11, PMO 83940/2. Road section west of Slemmestad, Røyken. Coll.: G. Henningsmoen, 1950. □D, G, H. Dorsal, lateral, and posterior view of pygidium, showing the well-defined posterior lobe and general convexity; ×18, PMO 121.610/2. Hagastrand, Asker. Coll. unknown, 1976. □E. Dorsal view of cephalon with prominent convexity; ×11, PMO 1298/2. Bjørkåsholmen, Asker. Coll.: L. Størmer, 1915. □F. Dorsal view of a well-preserved pygidium; ×22, PMO 61480/10. Bjørkåsholmen, Asker. Coll. unknown.

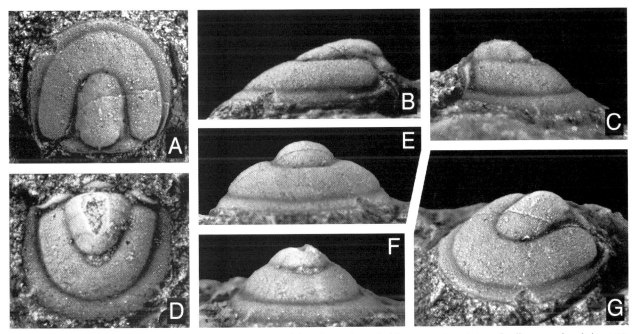

Fig. 18. Arthrorhachis mobergi (Tjernvik, 1956). □A, B, E, G. Dorsal, lateral, anterior and oblique anterolateral view of well-preserved cephalon; ×18, PMO 121.636. 32–40 cm above base of the formation, Øvre Øren, Modum. Coll.: J.O.R. Ebbestad, 1992-08-26. □C, D, F. Lateral, dorsal and posterior view of pygidium with faint rachial furrows; ×11, PMO 121.618/1. Bjørkåsholmen?, Asker. Coll.: F. Nikolaisen.

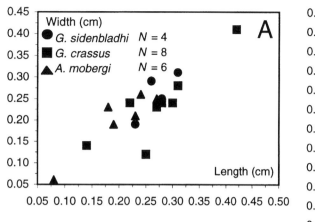

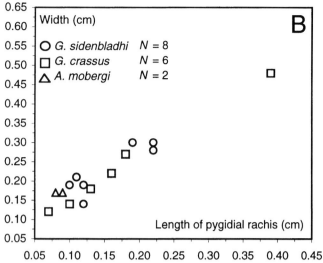

Fig. 19. Cephala and pygidia of agnostid species in the Bjørkåsholmen Formation. □A. Length (sag.) of cephala plotted against width (tr.) of cephala of *Geragnostus sidenbladhi*, *G. crassus* and *Arthrohrachis mobergi*. □B. Length of pygidial rachis (sag.) plotted against width (tr.) of pygidia of *Geragnostus sidenbladhi*, *G. crassus* and *Arthrohrachis mobergi*.

Discussion. – The species was redescribed and figured by Ahlberg (1992, pp. 560–562, Fig. 16a–i), and a redescription is not necessary here. The Norwegian material is scarce and mostly incomplete, but corresponds closely to the Swedish specimens, except that the anterior margin of the figured pygidium (Fig. 18C, D, F) curves more markedly downwards and forwards lateral to the dorsal furrow. The anterior rachial lobes of the pygidium are also more indistinct laterally than in the depicted Swedish material (Ahlberg 1992, p. 562, Fig. 16). Tables 5 and 6 give the measurements of the cephala and pygidia, respectively.

Table 5. Cephalic measurements of *Arthrorhachis mobergi*.

Specimen	A	B	B1	J	J3	K
1323/6	0.08	0.04	–	0.06	0.07	0.03
61479/11	0.24	0.15	0.21	0.26	0.26	0.09
63159/2	0.19	0.12	0.18	0.19	–	0.08
83885/2	0.27	0.18	0.25	0.25	0.25	0.08
84002/2	0.18	0.11	0.35	0.23	0.19	0.10
121.636	0.23	0.13	0.24	0.21	0.21	0.09

Table 6. Pygidial measurements of *Arthrorhachis mobergi*.

Specimen	X	Y	Z	Z1	W	W2
1323/7	0.06	0.09	0.15	0.13	0.17	0.16
121.618/1	0.07	0.08	0.14	0.12	0.17	0.13

Family Eulomidae Kobayashi, 1955

Subfamily Eulominae Kobayashi, 1955

Genus *Euloma* Angelin, 1854

Type species. – *Euloma laeve* Angelin, 1854, p. 61, Pl. 33:14a, b, from the Biozone of *Megistaspis estonica* in the Hunneberg Group (Arenig) at Berg in Östergötland, Sweden; subsequently designated by Vodges (1925, p. 100).

Discussion. – Sdzuy (1958), Courtessole & Pillet (1975), and Peng (1990) listed European and Asian species of this genus. *Proteuloma*, proposed as subgenus of *Euloma* by Sdzuy (1958) and Courtessole & Pillet (1975), is to be regarded as a separate genus, and its species are to be excluded from the above-mentioned listing. The affinities of the two genera were also discussed by Owens *et al.* (1982) and Shergold & Sdzuy (1984). The latter authors suggested restricting the Subfamily Eulominae to a generic group including *Euloma* Angelin, 1854, *Pareuloma* Rasetti, 1954, *Lateuloma* Dean, 1973, and *Proteuloma* Sdzuy, 1958. Based on new material from China and Kazakhstan (Peng 1990 and references therein), *Ketyna* Rozova, 1963, and *Archaeuloma* Li *in* Yin & Li, 1978, with weak glabellar furrows, should be included in the same subfamily. These are both pre-Tremadoc genera. Shergold & Sdzuy (1984, p. 84) suggested that some eulomid genera might be descended from species of *Ketyna*.

Euloma ornatum Angelin, 1854

Figs. 20–22

Synonymy. – □1854 *Euloma ornatum* n.sp. – Angelin, p. 92, Pl. 42:3, 3a–c. □1857 *Euloma ornatum* Angelin – Kjerulf, p. 93. □1865 *Euloma ornatum* Angelin – Kjerulf, p. 2. □1869 *Euloma ornatum* Angelin – Linnarsson, p. 72. □1882 *Euloma ornatum* Angelin – Brøgger, pp. 97–98, Pl. 3:5, 6. □1901 *Euloma ornatum* Angelin – Holm, p. 35, Fig. 29. □1906 *Euloma ornatum* Angelin – Moberg & Segerberg, p. 84, Pl. 4:41–44. □1940 *Euloma ornatum* Angelin – Lake, p. 302. □1956a *Euloma ornatum* Angelin – Tjernvik, p. 275, Text-fig. 45a, Pl. 9:4, 5. □1990 *Euloma ornatum* Angelin – Peng, p. 80.

Type material. – Angelin (1854, p. 92) reported occurrence of the species in the Bjørkåsholmen Formation at Hunneberg in Västergötland, Sweden, and from Oslo, Norway. The figured specimens (Angelin 1854, Pl. 42:3, 3a–c) are missing from the type collection at the Swedish Museum of Natural History, Stockholm, and it is therefore impossible to assert their actual localities or select a lectotype. Until such time as the originals are found or a neotype is selected, the Norwegian material must suffice to define the taxon.

Norwegian material. – Cranidia, pygidia and free cheeks occur abundantly in the formation across the Oslo Region. In addition, four nearly complete specimens and two hypostomes are known.

Remarks. – Angelin (1854) gave a short diagnosis of the species, while Moberg & Segerberg (1906) gave a limited description. Since then, little has been added to the knowledge of the species, and an emended diagnosis and description are given here.

Emended diagnosis. – Glabella with three pairs of distinct glabellar furrows. Palpebral lobes large, rising to near level of glabella. Anterior border convex (sag.), curving steeply downwards. Anterior border furrow with 9–12 distinct pits. Pygidium semielliptical, wider (tr.) than long (sag.), with three posterior diverging pleural ribs.

Emended description. – Sagittal length of cranidium half posterior width. Glabella convex (tr.), evenly rounded abrachially, leaving a faint sagittal line. It tapers forward, maximum width (tr.) of frontal glabellar lobe two-thirds of occipital ring width (tr.). Occipital ring tapers (tr.) behind 1L lobes, defined anteriorly by prominent occipital furrow. Occipital width (tr.) one-third of posterior width. Median occipital tubercle distinct. The 1S furrows are deep, converging rearwards from dorsal furrow at 150°, then geniculating, orienting the posterior part exsagittally. The 1L lobes are almost isolated posteriorly, well rounded, tapering towards dorsal furrow. The 2S furrows are deep, widest (sag.) marginally, converging

rearwards at 150°, not as deep adrachially as 1S furrows. The 2L lobes are rounded, slightly tapering marginally, separated from fixed cheek by faintly indicated dorsal furrows. The 3S furrows are small, distinct, situated opposite anterior extremity of palpebral lobes. Frontal glabellar lobe rounded, semicircular in outline, somewhat truncated anteriorly. Border furrow distinct, continuing into deep anterior pits anterolateral to glabella. Preglabellar furrow well defined, convex (tr.), deeper near anterior pits. Facial suture opisthoparian. Fixed cheek narrow (exsag.) posteriorly, the width (tr.) being twice the anterior width. Posterior margin curves slightly backwards towards genal angle. Posterior border furrow wide, U-shaped (sag.), and shallow. Palpebral lobes large, semicircular in outline, making up one-third of sagittal length. They rise steeply to near level of glabella, width (tr.) nearly two-thirds of posterior width, and are separated from the fixed cheeks by well-defined palpebral furrows. From anterior pits, low-relief eye ridges diverge posteriorly towards anterior part of palpebral furrows. Preglabellar field highly convex (sag.), flattening somewhat marginally, and curving steeply downwards. Anterior branches of facial suture diverging towards the somewhat semielliptical anterior margin, slightly pointed sagittally. Anterior border narrow, defining a less steeply curved terrace. Anterior border furrow distinct, but narrow, with 9–12 irregular spaced pits. Anterior arch wide and weakly curved. Test bears fine caeca on the preglabellar area (Fig. 20L), otherwise exoskeletal parts of cranidium carries fine granulation.

Free cheek elongate with long genal spines extending backwards as continuation of curved cephalic margin. Inner margin at the base of genal spine curves strongly backwards. Eye socle elevated, concave (tr.). Genal areas convex (exsag. and tr.) with convex (tr.), bevelled lateral borders defined by border furrows. Exoskeleton with fine caeca directed from eye socle to border furrow.

Hypostomes: The two Norwegian hypostomes (one depicted; Fig. 20H, K) resemble that described by Tjernvik (1956a, p. 275, Pl. 11:5). Outline slender and elongated. Width posterior of anterior wing, two-thirds of sagittal length. Anterior and posterior lobes convex (tr.), with a slightly concave transition. Elongated, paired maculae present. Border furrow well defined, with a narrow lateral border which is well rounded posteriorly.

Thoracic region with twelve segments, tapering backwards (Fig. 20F, G, I). Half-rings and rachial rings highly convex (tr.) with about the same width (sag.). Rachial ring parallel-sided (tr.), curving forward towards the rachial furrow. Pleural furrows wide (sag.), lying between low anterior and posterior flanges. They are spindle-shaped at lateral extremities, curving backwards.

Pygidium: Sagittal length of pygidium somewhat more than one-third anterior width. Rachis tapers backwards, making up 80% of sagittal length, carrying three rachial

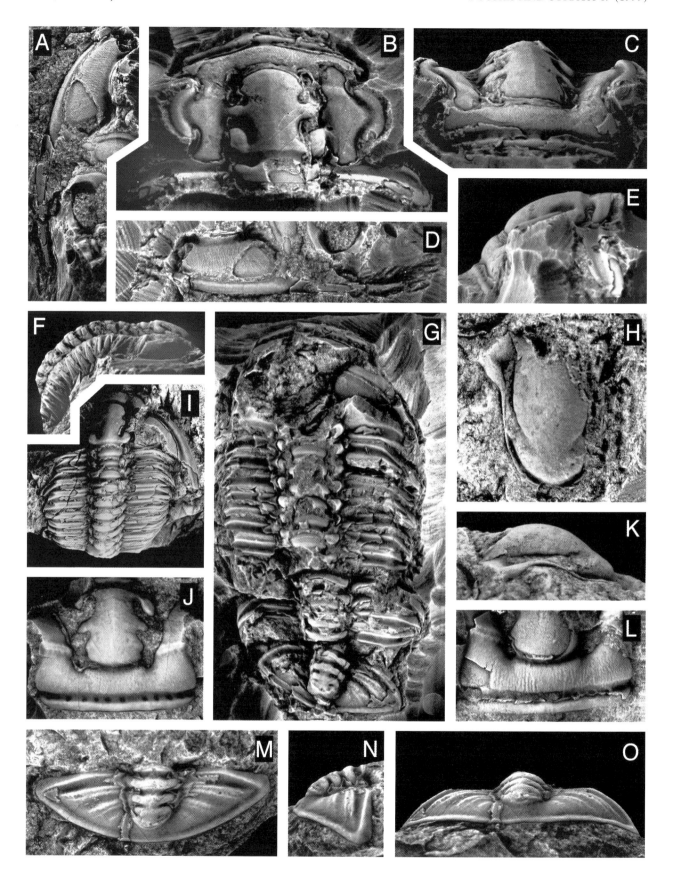

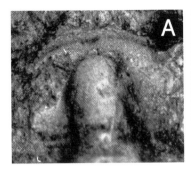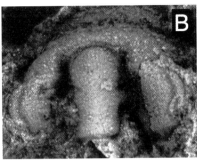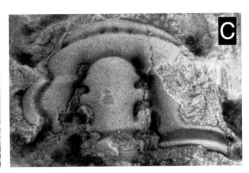

Fig. 21. Euloma ornatum Angelin, 1854. □A. Dorsal view of early meraspid stage, showing anterior border and right palpebral lobe; ×20, PMO 136.022/1. Found 81–92 cm above base of the formation at the western side of Jarenvannet, Hadeland. Coll.: J.O.R. Ebbestad, 1992-09-18. □B. Dorsal view of meraspid stage, showing glabellar furrows and weakly developed palpebral lobes ; ×20, PMO 60324/2. Ramtonholmen, Røyken. Coll. unknown, 1902-08-19. □C. Dorsal view of small cranidium; ×11, PMO 60321/3. Vestfossen, Øvre Eiker. Coll.: W.C. Brøgger, 1879.

rings, the posterior one less distinct. Terminal rachial piece rounded, obtuse posteriorly, extending steeply, slightly concave (sag.), down to posterior margin. Outline of posterior margin semielliptical. Posterior border narrow, tapering adrachially. Anterior margin transverse abrachial for about half the posterior width before curving steeply down laterally. Pleural regions convex (sag. and tr.), having three pairs of pleural ribs with a shallow central furrow. Posterior ribs with low relief.

Ontogeny. – A series of small growth stages is figured (Fig. 21A–C), showing a gradational development of the cranidium of *Euloma ornatum*. The two smallest stages (Fig. 21A, B) show the anterior border, palpebral lobes, and glabellar furrows further developed. The last stage (Fig. 21C) measures 0.33 cm (sag.), and already the adult characters are fully developed. From the smallest cranidium found (0.15 cm long) to the largest (1.63 cm long), there is an increase in size of nearly 11 times.

The growth series show that the major morphological changes appear very early in the meraspid period. During growth the glabella becomes proportionally shorter and wider, tapering forward, and the fixed cheek becomes narrower (tr.). Finally the anterior border, border furrow and palpebral lobes develop, establishing the adult proportions within the following moulting stages.

Discussion. – A short diagnosis and description of the Arenig type species *Euloma laeve* were given by Tjernvik (1956a, p. 274, Text-fig. 44B, Pl. 11:1–3), based on comparison with the present Tremadoc species. The latter differ from the type species mainly in having a proportionally longer glabella with three glabellar furrows instead of two, smaller palpebral lobes, a more convex anterior border, and a smooth semielliptical posterior pygidial margin (see Fig. 22 for reconstruction). In other respects, the two species are quite similar. The segmented pygidium with three distinct pleural ribs is a characteristic and shared feature (Peng 1990, p. 80).

This is also seen in *E. filacovi* Bergeron, 1889, of the upper Tremadoc–Arenig series of Montagne Noir, France. It was revised by Courtessole & Pillet (1975, p. 256), and a subspecies, *E. filacovi mourezense*, from the same area, was described by Berard (1986). *E. filacovi* differs from *E. ornatum* mainly in the distinction of 3S furrows, preglabellar furrow, and the proportionally longer anterior branches of the facial suture. *E. orientale* Liu *in* Zhou *et al.*, 1977, from the lower part of the Madaoyu Formation, upper Tremadoc of South China, resembles the broadly coeval *E. laeve*, *E. filacovi* and *E. ornatum*. Peng (1990) showed that the Chinese species has a generally proportionally longer and less segmented pygidium, and compared to *E. ornatum* the glabella appears to be proportionally longer and its anterior border is relatively narrower (sag.). The palpebral lobe furrow of the Chinese form is less distinct, and the strongly granulated test is strikingly different from that of *E. ornatum*.

In the uppermost part of the underlying Alum Shale Formation, formerly Ceratopyge Shale (3aγ), in the Oslo Region, rare specimens of *E. ornatum* occur in limestone

Fig. 20. Euloma ornatum Angelin, 1854. □A, D. Dorsal and lateral view of free cheek, showing the long genal spine and prominent eye socle; ×3, PMO 1382, Bjørkåsholmen, Asker. Coll. unknown, 1915. □B, C, E. Dorsal, anterior and lateral view of well-preserved cranidium; ×3, PMO 121.565/2. Bjørkåsholmen, Asker. Coll.: F. Nikolaisen, 1960-05-01. □F, I. Lateral and dorsal view of nearly complete specimen, carrying twelve thoracic segments; ×3, PMO H1322. Vestfossen, Øvre Eiker. Coll.: W.C. Brøgger, 1879. □G. Nearly complete moulting stage, showing thoracic region carrying eleven thoracic segments; ×2, PMO 60325/3. Bjørkåsholmen, Asker. Coll.: J. Kiær, 1922-09-25. □H, K. Dorsal and lateral view of hypostome; ×8, PMO 136.082. Bjørkåsholmen, Asker. Coll.: B. Funke, 1978-11-04. □J. Oblique anterior view of cranidium, showing anterior border furrow with pits and the eye ridges; ×3, PMO 84049/3. Bjørkåsholmen, Asker. Coll.: G. Henningsmoen, 1958-11-16. □L. Oblique anterior view of cranidium retaining most of the exoskeleton; ×3, PMO 136.076. Bjørkåsholmen, Asker. Coll.: F. Nikolaisen, 1960-05-01. □M, N, O. Dorsal, lateral and posterior view of pygidium; ×3, PMO 121.604. Bjørkåsholmen, Asker. Coll.: F. Nikolaisen, 1960-05-01.

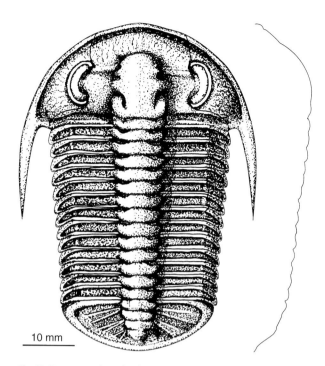

Fig. 22. Reconstruction of *Euloma ornatum* Angelin, 1854.

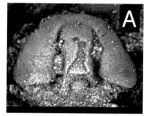

Fig. 23. Hospes? sp. □A. Dorsal view of cephalon; ×20, PMO 61480/19. Bjørkåsholmen, Asker. Coll. unknown. □B. Dorsal view of cephalon, showing three faint glabellar furrows; ×20, PMO 61480/12. Bjørkåsholmen, Asker. Coll. unknown.

Hospes? sp.

Fig. 23

Material. – Two cephala, PMO 61480/12 and 61480/19, from the level above the dark limestone nodules at Bjørkåsholmen, Slemmestad.

Description. – Glabella rectangular, making up 80% of total cephalic length (sag.) and 45% of maximum cephalic width (tr.), three pairs of glabellar furrows faintly indicated laterally, evenly spaced along glabella. Occipital ring distinct, dorsal furrows well defined, anterior glabellar lobe bluntly terminated at indistinct preglabellar furrow. Preglabellar field short (sag.). Cephalon tapering anteriorly, anterior margin transverse, posterior border furrows weakly indicated.

Discussion. – The rounded triangular cephalon, the narrow glabella with the bluntly terminated front, the indistinct preglabellar furrow and the short preglabellar field suggest similarity with *Hospes clonograpti*, and the Norwegian specimens are tentatively referred to this genus. The Norwegian material, however, shows prominent occipital rings, distinct dorsal furrows and faint traces of glabellar furrows that are not clearly seen in the type species. The posterior border furrow seen in *H. clonograpti* is only faintly indicated on the Norwegian material.

nodules and calcareous beds. These resemble specimens from the type horizon, except that in the cranidia available (PMO 689/1 and PMO 20036), both equal in size to the specimen in Fig. 21C, the palpebral furrows are less distinct. This might be a result of preservation.

Family Conocoryphidae Angelin, 1854

Genus *Hospes* Stubblefield *in* Stubblefield & Bulman, 1927

Type species. – *Hospes clonograpti* Stubblefield *in* Stubblefield & Bulman, 1927, pp. 128–130, Pl. 4:2, from the Shineton Shales (Tremadoc) at Cherme's Dingle near Garmston, Shropshire, England; by original designation of Stubblefield *in* Stubblefield & Bulman (1927, p. 128).

Discussion. – This rare genus occurs in Upper Cambrian and Lower Ordovician strata in England and China, respectively. Species included are *Hospes clonograpti* Stubblefield *in* Stubblefield & Bulman, 1927, from Shropshire, England, and two Chinese species from the Anhui district, *H. carinatus* Qian, 1985 and *H. transversa* Qian, 1985. The Norwegian specimens described below may prove to be the stratigraphically youngest representatives of this genus.

Family Shumardiidae Lake, 1907

Genus *Shumardia* Billings, 1862

Type species. – *Shumardia granulosa* Billings, 1862, pp. 92–93, Fig. 83, from the *Shumardia* limestone (lower Llanvirn) at Lévis in Quebec, Canada; by original designation of Billings (1862, p. 92).

Subgenus *Shumardia* (*Conophrys*) Callaway, 1877

Type species. – *Conophrys salopiensis* Callaway, 1877, p. 677, Pl. 24:7, from the Shineton Shale Formation (Tremadoc) in South Shropshire, England; by subsequent monotypy.

Discussion. – Considering the type species of *Shumardia* Billings, 1862, *Conophrys* Callaway, 1877, *Kweichowiella* Chang, 1964 (*in* Chang *et al.* 1964), and *Shumardia* (*Shumardella*) Přibyl & Vaněk, 1980, four *Shumardia*-like taxa are easily definable (Fortey & Owens 1987). However, taking into account the great morphological range of described shumardiid species (Fortey 1980 and references therein), these four taxa are best recognized as subgenera of *Shumardia* (see Fortey & Owens 1987, pp. 119–120).

The subgenus *S.* (*Conophrys*) is distinguished mainly by the small to moderate-sized anterolateral glabellar lobes which are defined inside the cephalic margin. They are generally larger, more swollen, in the other subgenera. Peng (1990) adopted the proposed diagnosis of *S.* (*Conophrys*) with the modification that the thoracic segments may vary in number from five to seven. Thoracic segments are unknown from the Norwegian material of *S.* (*C.*) *pusilla*. However, the specimens conform well with the subgeneric diagnosis given by Fortey & Owens (1987), except that the pygidial rachis does not extend to the border, a feature shared by the Swedish species *S.* (*C.*) *nericiensis* Wiman, 1905.

Dean (1973) proposed reviving *Conophrys* Callaway, 1877, to a generic level, but Fortey (1980) concluded that in spite of the differences between their type species, *Shumardia* and *Conophrys* could not equivocally be divided into two generic groups because their species intergrade. This is evident in *S. minaretta* Fortey, 1980, which has a *Shumardia*-like cephalon and a *Conophrys*-like pygidium.

Shumardia (*Conophrys*) *pusilla* (Sars, 1835)

Figs. 24, 25A

Synonymy. – □1835 *Battus pusillus* n.sp. – Sars, p. 334, Pl. 8:2a, b. □1838 *Trilobitus pusillus* (Sars) – Boeck, p. 144. □1857 *Battus pusillus* (Sars) – Kjerulf, p. 144. □1865 *Agnostus* (*Olenus?*) *pusillus* (Sars) – Kjerulf, p. 2, Fig. 10. □1882 *Conophrys pusilla* (Sars) – Brøgger, p. 125, Pl. 12:9. □1890 *Shumardia pusilla* (Sars) – Moberg, p. 4. □1906 *Shumardia pusilla* (Sars) – Moberg & Segerberg, p. 80, Pl. 4:10–12. □non 1907 *Shumardia pusilla* (Sars) – Lake, p. 40, Pl. 3:18–20. □1920 *Shumardia pusilla*(?) (Sars) – Størmer, p. 13, Pl. 2:2. □non 1926 *Shumardia pusilla* (Sars) – Stubblefield, p. 345, Pl. 14–16. □1940 *Shumardia pusilla* (Sars) – Størmer, p. 128, Pl. 1:13. □1952 *Shumar-dia pusilla* (Sars) – Hutchinson, p. 95, Pl. 5:10. □1959 *Shumardia pusilla* (Sars) – Harrington *et al. in* Moore, p. 0245, Fig. 183. □non 1973 *Shumardia pusilla* (Sars) – Bulman & Rushton, Pl. 6:1–4. □1978 *Shumardia pusilla* (Sars) – Courtessole & Pillet, pp. 179–180, Pl. 1:1–13. □1987 *Shumardia* (*Conophrys*) *pusilla* (Sars) – Fortey & Owens, p. 120. □1990 *Shumardia* (*Conophrys*) *pusilla* (Sars) – Peng, p. 16. □1991 *Shumardia* (*Conophrys*) *pusilla* (Sars) – Fortey & Owens, p. 461, Fig. 8s–w.

Lectotype. – A cephalon (PMO 56420) from the uppermost part of the Alum Shale Formation (Tremadoc), formerly Ceratopyge shale (3aβ), at Sofienberg near Drammensveien, Oslo, Norway. Selected and figured by Størmer (1940, p. 128, Pl. 1:13). Illustrated also by Fortey & Owens (1991, Fig. 8u).

Material. – Twenty-two cranidia, three pygidia. Largest cranidium 0.16 cm long (Fig. 24E), smallest cranidium 0.05 cm long.

Remarks. – A diagnosis has never been given for this species. All former descriptions are also inadequate, since it was generally accepted that British and Norwegian material of *Shumardia pusilla* were conspecific. Fortey & Owens (1991), however, proved that this was not the case. A diagnosis and a new description are provided here.

Diagnosis. – *Shumardia* (*Conophrys*) with well-defined frontal glabellar lobe. Dorsal furrow deep, posterior of anterolateral lobes. Anterior margin thin (horizontally) and narrow (sag.), curving slightly backwards sagittally. Pygidium with four rachial rings, rachis only two-thirds of sagittal length.

Emended description. – Cephalon convex, sagittal length half posterior width. Glabella occupies 90% of sagittal length and one-third of posterior width. Posterior part convex while anterior part expands (tr.), being nearly one-third wider (tr.) than the posterior part of glabella. Occipital ring convex (tr.) tapering distally, with small, indistinct node at posterior extremity (Fig. 24C, E), separated anteriorly by distinct, straight occipital furrow. The 1S furrows are wide (exsag.) marginally, not very deep adrachially, converging obliquely backwards (Fig. 24D, H). The 1L lobes are small, almost isolated posteriorly. The 2S furrows converge obliquely forward, tapering as they outline the anterolateral lobes. Frontal glabellar lobe slanting forward to preglabellar furrow. Dorsal furrows as deep indentations posterior to anterolateral lobes, shallowing as they curve forward around the lobes and confluent with the preglabellar furrow, the latter being semielliptical in outline. Cheeks convex, sloping nearly vertically distally. Semicircular in outline, merging with a transverse anterior margin, curving slightly inwards sagittally. Posterior margin transverse or diverging slightly backwards, projecting into small, but distinct, genal

bellar area. Much of this variation may be due to preservation, the cheeks flexing outwards during compaction of the rock, but some morphological variation clearly exists within the species.

Frank Nikolaisen, Oslo (personal communication, 1993), kindly drew my attention to a specimen of a *Shumardia* species (PMO 136.086; Fig. 24K) found in the Bjørkåsholmen Formation at the type locality at Slemmestad, Norway. This specimen is clearly different from *S. (C.) pusilla*. The posterior part of the glabella is wider (tr.), without prominent transition to the large, swollen anterior part. The anterolateral lobes are small, not well defined; the dorsal furrow merges with a weak, well-rounded preglabellar furrow; and spines are not as well-developed at the genal angle. The specimen has some affinities towards *S. (C.) nericiensis* Wiman, 1905, but based on one specimen only, it is not assigned to any species here.

Family Olenidae Burmeister, 1843

Subfamily Hypermecaspidinae Harrington & Leanza, 1957

Genus *Hypermecaspis* Harrington & Leanza, 1957

Type species. – *Hypermecaspis inermis* Harrington & Leanza, 1957, pp. 121–135, Fig. 48:1–2, from the lower Tremadoc *Kainella meridionalis* Biozone of Salta, Argentina; by original designation of Harrington & Leanza (1957, p. 121).

Discussion. – Following Henningsmoen (1959) and Fortey (1974), the Hypermecaspidinae is regarded a subfamily of the Olenidae. To the modification of the subfamily given by Fortey (1974) can be added that the maximum width (tr.) of the pygidium is situated opposite or posterior to half the maximum length (sag.) of the pygidium.

Under the present understanding, the genus *Hypermecaspis* includes fourteen species: *H. rugosa* Brøgger, 1882; *H. armata* Harrington & Leanza, 1957; *H. bulmani* Harrington & Leanza, 1957; *H. inermis* Harrington & Leanza, 1957; *H. cf. bulmani* Whittington, 1965; *H. boliviensis* Branisa, 1965; *H. kolouros* Ross, 1970; *H. alveus* (Fortey, 1974); *H. venulosa* Fortey, 1974; *H. latigena* Fortey, 1974; *H. brevifrons* Fortey, 1974; *H. branisai* Přibyl & Vaněk, 1980; *H. venerabilis* Fortey & Owens, 1978; and *H. minitis* Henderson, 1983. See Fig. 26 for distribution.

Tropidopyge alveus Fortey, 1974, from the Middle Ordovician of Svalbard, is here assigned to *Hypermecaspis*. The diverging anterior facial sutures and a conspicuous preglabellar field are generally similar in *H. latigena*.

Originally, characters of the palpebral lobes and the pygidium were used for the generic separation (Fortey 1974). The palpebral lobes are positioned far away from the glabella, but this can also be observed in *H. armata* Harrington & Leanza, 1957, from Argentina. The pygidium resembles that of *T. broeggeri* (Moberg & Segerberg, 1906), but it also closely resembles that of *H. inermis* Harrington & Leanza, 1957, from Argentina, both in outline and distinction of the pleural and interpleural furrows. The rachis is longer in *H. inermis*, but in species such as *H. latigena* Fortey, 1974, and *H. brevifrons* Fortey, 1974, from Svalbard the rachis is equally short.

The pygidium of *H. inflecta* (Harrington & Leanza, 1957) differs from that of other species of *Hypermecaspis* in having a less tapering rachis, bluntly rounded posteriorly, and a semicircular pygidium. Harrington & Leanza (1957, p. 126) described a short postrachial ridge, but this is not evident on the illustrations. It is probably not congeneric and is here excluded from the *Hypermecaspis*. The species is tentatively assigned to *Cermatops* Shergold, 1980. The pygidia figured by Harrington & Leanza (1957, p. 124, Fig. 48:1, 2) are strikingly similar to the pygidium of *Cermatops thalastus* Jell et al. (1991, Fig. 7F–H).

The occurrence of *Hypermecaspis inflecta* Harrington & Leanza, 1957, identified by Gjessing (1976a, unpublished data referred to by Owen et al. 1990, p. 7, as *Høypermecaspis* [sic] *inflecta*) from the upper part of the Alum Shale Formation in the Oslo Region, is here considered incorrect. The Norwegian specimen has fewer rachial rings and disitinct pleural fields. As for *H. inflecta* the Norwegian specimen has no affinity to *Hypermecaspis*.

Přibyl & Vaněk (1980) established the subgenus *Hypermecaspis* (*Spitsbergaspis*) for the three species of *Hypermecaspis* described from Svalbard by Fortey (1974). The remaining *Hypermecaspis* species were assigned to the subgenus *H. (Hypermecaspis)* Harrington & Leanza, 1957. The main reasons for this distinction were the presence of caeca on the preglabellar fields and the divergent facial sutures on the former species. Divergent facial sutures are also present in *H. venerabilis* (Fortey & Owens, 1978), but the latter species lacks the caeca. This suggests that the two characters are not suitable for subgeneric separation.

Harrington & Leanza (1957), Henningsmoen (1959), Harrington et al. (in Moore 1959), and Fortey (1974) have suggested that the type species *Tropidopyge broeggeri* (Moberg & Segerberg, 1906), known only from a few pygidia in Norway and Sweden, and *Hypermecaspis rugosa* Brøgger, 1882, known only from two cranidia in Norway, may be conspecific. Even though the morphological variations within the two groups are coherent, such a procedure would be premature, since a definite association of *Hypermecaspis rugosa* Brøgger, 1882, and *Tropidopyge broeggeri* (Moberg & Segerberg, 1906) cannot be established at the moment (see discussion below on *Hypermecaspis rugosa* and *Tropidopyge broeggeri*). However, if the

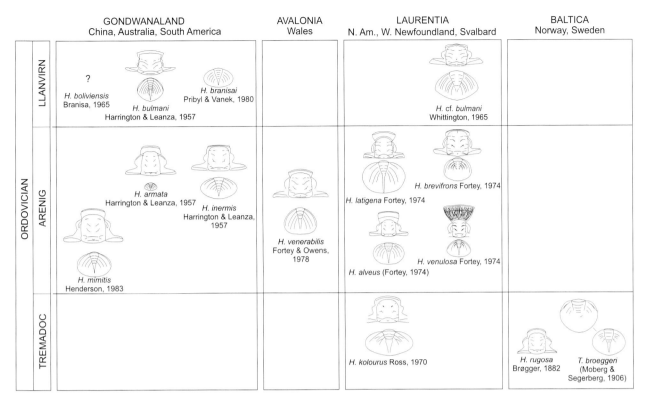

Fig. 26. Distribution and comparison of species attributed to *Hypermecaspis*, including *Tropidopyge broeggeri.*

two species were to be synonymized, *Hypermecaspis* would become a junior synonym of *Tropidopyge*. Given the concepts of *Hypermecaspis* and *Tropidopyge* outlined herein, the latter genus contains only the type species.

There is a great deal of morphological variation in the basic pygidial features within the two groups. The type pygidium of *Tropidopyge* resembles that of *Hypermecaspis alveus* and *H. inermis*, but the latter two lack the antero-lateral marginal brim at maximum width (tr.) of the pygidium and the prominent articulating facets. The rachis of *H. inermis* carries an additional rachial ring. An anterolateral brim and articulating facets are seen to some extent in *H. venulosa* and, possibly, *H. bulmani*. The pygidium of these two species has, in turn, a more fan-shaped outline. The rachis of the type specimen carries four rachial rings, as does *H. alveus, H. latigena* and *H. mimitis.*The postrachial ridge are proportionally the same size in these species. In other species, such as *H. kolourus, H. latigena* and *H. venerabilis,* the ridge is longer (sag.) than the rachis, and the rachis carries three rachial rings.

The cranidia of *Hypermecaspis* resemble *Parabolinella* Brøgger, 1882, the proposed ancestral genus (Henningsmoen 1959). The two genera differ mainly in the complexity of the glabellar segmentation (Harrington & Leanza 1957).

Hypermecaspis rugosa Brøgger, 1882

Figs. 27, 28

Synonymy. – □1882 *Parabolina rugosa* n.sp. – Brøgger, p. 104, Pl. 3:3. □*non* 1896 *Parabolinella* sp. (Brøgger) – Crosfield & Skeat, p. 537, Pl. 26:11, 12. □1906 *Parabolinella rugosa* (Brøgger) – Moberg & Segerberg, p. 82. □*non* 1913 *Parabolinella rugosa* (Brøgger), var. – Lake, p. 67, Pl. 7:3. □1951 *Parabolinella rugosa* (Brøgger) – Shaw, p. 103. □1957 *Parabolinella rugosa* (Brøgger) – Henningsmoen, p. 137, Pl. 12:9. □1957 *Hypermecaspis rugosa* (Brøgger) – Harrington & Leanza, pp. 121, 123. □1959 *Hypermecaspis rugosa* (Brøgger) – Harrington *et al. in* Moore, p. O270. □1959 *Hypermecaspis rugosa* (Brøgger) – Henningsmoen, p. 161. □1974 'Parabolinella'*rugosa* (Brøgger) – Fortey, p. 51.

Holotype. – An incomplete cranidium from a dark lime-stone nodule near the base of the Bjørkåsholmen Formation at Vestfossen, Øvre Eiker, Norway. Described and illustrated by Brøgger (1882, pp. 104–105, Pl. 3:3). By monotypy. The specimen has not been located in the collections at the Paleontological Museum, Oslo, and may be lost. A cranidium (PMO 1267/1) from the same horizon at Bjørkåsholmen, Asker, is probably conspecific with the holotype but is presently not selected as neotype.

Fig. 27. Hypermecaspis rugosa Brøgger, 1882. Dorsal view of incomplete cranidium; ×9, PMO 1267/1. Dark limestone nodule 25–30 cm above the base of the formation, Bjørkåsholmen, Asker. Coll. unknown, 1915. Figured by Henningsmoen (1957, Pl. 12:9).

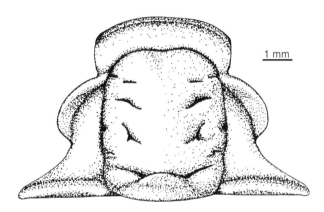

Fig. 28. Reconstruction of the cranidium of *Hypermecaspis rugosa* Brøgger, 1882, based on the figured specimen (Fig. 27, PMO 1267/1) and the drawing of the type (Brøgger 1882, Pl. 3:3).

Material. – A cranidium, 0.5 cm long (sag.), and a drawing (by Brøgger 1882, Pl. 3:3) of the holotype, 0.7 cm long (sag.).

Emended description. – Brøgger (1882, p. 104) gave a diagnosis and a description. A short description was also given by Henningsmoen (1957). A new, more extensive description is provided here.

Sagittal length of cranidium nearly two-thirds of posterior width. Glabella slightly convex (tr.), sloping near vertical distally, occupying 80% of sagittal length. It tapers only slightly anteriorly. Occipital ring damaged on specimen but shows a crescentic median part and elongated triangular lateral portions. Median node not observed. Occipital furrow with anteriorly convex median part prolonged laterally by anteriorly diverging branches, not reaching dorsal furrow. The 1S furrows are merely faint

indications laterally, glabella curving slightly inwards at this point. The 2S furrows are distinct, converging rearwards from dorsal furrow at 150°, bifurcating laterally. Marginally, opposite bifurcation, distinct furrows appear in association with 2S furrows but do not merge. The 3S furrows are distinct, directed adrachially, then geniculating rearwards from dorsal furrow at 150°. The 4S and 5S furrows are short (tr.) less distinct, almost transverse. The 4S furrows are situated adrachially just posterior to 5S furrows that are reaching dorsal furrow. Frontal glabellar lobe semielliptical, curving slightly backwards sagittally. Fixed cheeks wide anteriorly with almost transverse posterior margin and shallow, wide (sag.) posterior border furrow. Palpebral lobes incomplete on specimen, but must have been large, situated between levels of 5S glabellar furrows and 2S glabellar furrows. Eye ridges short, curving obliquely backwards from opposite 5S glabellar furrows. Fixed cheek narrow (tr.) at this point. Posterior parts of anterior facial suture diverge outwards, then converge anteriorly to merge with anterior margin. Width of poorly preserved preglabellar area, nearly half posterior width, curving slightly downwards. Anterior border poorly defined.

Fig. 28 presents a reconstruction of *H. rugosa*.

Discussion. – The illustrated cranidium (Fig. 27, PMO 1267/1) and the drawing of the holotype (Brøgger 1882, Pl. 3:3) are similar in size, morphology and stratigraphical position, and are here treated as conspecific. As noted by Henningsmoen (1957, p. 137), the S2 furrows curve more markedly backwards on the drawing of the holotype.

H. rugosa resembles closely *H. inermis* Harrington & Leanza, 1957, *H. brevifrons* Fortey, 1974 , *H. venerabilis* Fortey & Owens, 1978, and *H. mimitis* Henderson, 1983, in the general outline of the cranidium. The differences lie mostly in the definition of the fixed cheeks and width of preglabellar field. *H. alveus* (Fortey, 1974) *H. latigena* Fortey, 1974, and *H. venulosa* Fortey, 1974, share a distinct expanded preglabellar field with divergent facial sutures and caeca. Caeca are also present in *H. brevifrons* Fortey, 1974, and *H. mimitis* Henderson, 1983, both having divergent facial sutures. The latter feature is also present in *H. rugosa* Brøgger, 1882, and *H. venerabilis* Fortey & Owens, 1978, but both lack the caeca. Species like *H. armata* Harrington & Leanza, 1957, and *H. inermis* Harrington & Leanza, 1957, have convergent and straight facial sutures, respectively and both lack the caeca.

Harrington & Leanza (1957), Henningsmoen (1959), Harrington *et al.* (*in* Moore 1959), and Fortey (1974) have suggested that *Tropidopyge broeggeri* (Moberg & Segerberg, 1906), known only from a few pygidia in Norway and Sweden, and *Hypermecaspis rugosa* Brøgger, 1882, known only from two cranidia in Norway, may be conspecific. *H. rugosa* occurs in dark limestone nodules at the base of the Bjørkåsholmen Formation, associated with

Bienvillia angelini. Of the two specimens of *T. broeggeri* included in morphotype 1 (see discussion of *T. broggeri*) the Norwegian specimen is from the same level as *H. rugosa*, and the lectotype probably is so, too. However, no cranidia of *H. rugosa* are found above the level of the dark limestone nodules in the Bjørkåsholmen Formation, to where morphotype 2 of *T. broeggeri* is probably confined. Since the material of *H. rugosa* is too limited to show its pattern of occurrence relative to *T. broeggeri*, the two species are for the time being kept separate.

Genus *Tropidopyge* Harrington & Kay, 1951

Type species. – *Dicellocephalus bröggeri* Moberg & Segerberg, 1906, p. 87, Pl. 5:7, 8, from the Biozone of *Apatokephalus serratus*, Bjørkåsholmen Formation (upper Tremadoc), at Ottenby on Öland, Sweden; by original designation of Harrington & Kay (1951, p. 663).

Remarks. – Until a positive match between a pygidium of the type species and a cranidium is found, the species must be diagnosed based on the pygidium alone. The original diagnosis by Harrington & Kay (1951) is here sligthly modified.

Emended diagnosis. – Pygidium elliptical, wider than long, maximum width (tr.) at mid-sagittal length. Pleural fields transversely convex, sagittally concave in transition to border. Pleural furrows indistinct. Rachis tapering posteriorly. Terminal piece indistinct, continuing posteriorly into narrow postrachial ridge almost reaching posterior border.

Discussion. – The Colombian species *T. stenorachis* Harrington & Kay, 1951, is here considered not to be congeneric, following Henningsmoen (1959) and Peng (1992). Its affinity may lie within the Ogygiocaridinae. Furthermore, *T. tevipis* Petrunina, 1973, *T. sibiricus* Petrunina, 1973, *T. laevis* Ergaliev, 1980, and *T. huanensis* Peng, 1992, are also excluded from the genus. They are considered to represent a separate entity. Based on cranidial features, Peng (1992) found that these species were closer related to the Subfamily Macropyginae of the Family Ceratopygidae. The pygidia of these species bear some resemblance to that of *T. broeggeri*, especially in having a postrachial ridge. However, they differ markedly in being semicircular in outline and having their maximum width (tr.) distinctly closer to the anterior margin. *Tropidopyge* has an elliptical pygidium, with the maximum width (tr.) situated midway or posterior to the midway sagittal length of the pygidium. The excluded species may tentatively be assigned to *Macropyge* s.l.

Tropidopyge broeggeri (Moberg & Segerberg, 1906)

Figs. 29, 30

Synonymy. – ☐1906 *Dicellocephalus bröggeri* n.sp. – Moberg & Segerberg, p. 87, Pl. 5:7, 8. ☐1914 *Platycolpus*? *bröggeri* (Moberg & Segerberg) – Walcott, p. 349. ☐1951 *Tropidopyge bröggeri* (Moberg & Segerberg) – Harrington & Kay, p. 663. ☐1957 *Tropidopyge broeggeri* (Moberg & Segerberg) – Harrington & Leanza, p. 120. ☐1959 *Tropidopyge broeggeri* (Moberg & Segerberg) – Harrington *et al.* in Moore, p. O270, Fig. 200. ☐1959 *Tropidopyge broeggeri* (Moberg & Segerberg) – Henningsmoen, p. 159, Pl. 1:5–7. ☐1959 *Tropidopyge broeggeri* (Moberg & Segerberg) – Harrington *et al.* in Moore, p. O270, Fig. 200. ☐1974 *Tropidopyge broeggeri* (Moberg & Segerberg) – Fortey, pp. 51, 53, Pl. 16:6. ☐1992 *Tropidopyge broeggeri* (Moberg & Segerberg) – Peng, pp. 94–95.

Lectotype. – A pygidium (LO 1836) from the Biozone of *Apatokephalus serratus*, Bjørkåsholmen Formation, at Ottenby on Öland, Sweden. Illustrated by Moberg & Segerberg (1906, Pl. 5:7). Selected by Harrington & Kay (1951, p. 663). Illustrated also by Fortey (1974, Pl. 16:6).

Norwegian material. – Three pygidia: PMO 84041, PMO 756 (incomplete) and PMO 69567 with counterpart PMO 69568.

Diagnosis. – As for the genus.

Emended description. – The species was described by Moberg & Segerberg (1906, pp. 87–88), and a short description of the Norwegian material was also provided by Henningsmoen (1959, p. 160). An improved description based on new material is presented here.

Maximum length (sag.) of pygidium 80% of maximum width (tr.). Rachis, including posterior transition to postrachial ridge, making up about 60% of sagittal length, tapering backwards to half its anterior width (tr.). It carries four rachial rings, rachial furrows attenuating distally. Terminal rachial lobe indistinct, convex transversely, projecting obliquely downwards posteriorly into narrow postrachial ridge without distinct transition to pleural fields. Pleural fields broad, transversely convex, sagittally concave. Three pleurae present, curving obliquely backwards with an indistinct transition to marginal parts of pleural region and border. The two anterior pleural furrows distinct proximal to rachis, posterior pleural furrows merely faint indications near dorsal furrow. Interpleural furrows faintly indicated. Pattern of separated, rarely branching, terrace ridges covering entire pleural region. Pygidial border wide, concave anterolaterally, forming marginal rims at maximum width (tr.). Margin semicircular posteriorly, semielliptical laterally, changing curvature at maximum width, which lies at mid-sagittal length. Anterior margin transverse close to rachis, curved lateral

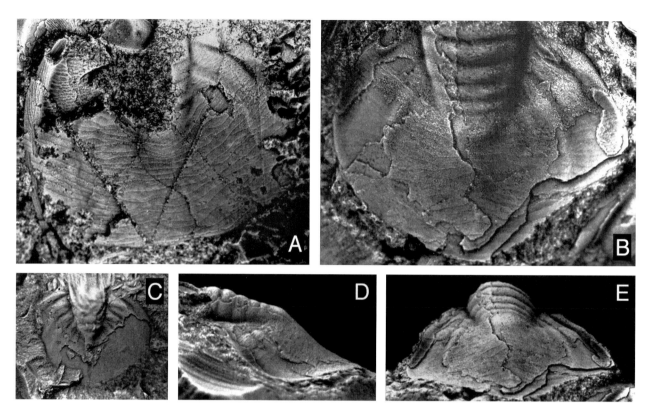

Fig. 29. Tropidopyge broeggeri (Moberg & Segerberg, 1906). □A. Dorsal view of pygidium, showing terrace ridges; ×9, PMO 756/1. Stensberggaten, Oslo. Coll.: J. Kiær. Figured by Henningsmoen (1959, Pl. 1:7). □B, D, E. Dorsal, lateral and posterior view of pygidium; B ×8; D, E ×6, PMO 84041/2. Bjørkås-holmen, Asker. Coll.: G. Henningsmoen, 1958-11-16. □C. Dorsal view of latex cast of small pygidium showing postrachial ridge; ×4, PMO 69568. Found in a dark limestone nodule 18–20 cm above the base of the formation at Sjøstrand, Asker. Coll.: G. Henningsmoen, 1958. Figured by Henningsmoen (1959, Pl. 1:6).

parts occupied by large articulating facets. Pygidial dou-blure wide.

Discussion. – The type material from Sweden (Moberg & Segerberg 1906, Pl. 5:7, 8, the latter rephotographed by Fortey 1974, Pl. 16:6) has been compared with the Norwe-gian material, and two morphotypes are recognized (Fig. 30, Table 7).

The first morphotype, based on the lectotype, pos-sesses a narrow postrachial ridge and pleural fields that are moderately inclined (Figs. 29C, 30A, C, E). The Nor-wegian specimen is equal in size to the lectotype and occurs in the dark limestone nodules near the base of the Bjørkåsholmen Formation. The dominant fossil in these

nodules is *Bienvillia angelini*. The level of the lectotype is unknown, but it is also associated with numerous speci-mens of *B. angelini*, suggesting that the sample might originate from the same stratigraphical level as the Nor-wegian one. Morphotype 1 resembles *Hypermecaspis inermis* Harrington & Leanza, 1957, except that the rachis is shorter, the outline is more semielliptical, and the body is smaller. It is much like *Hypermecaspis venulosa* (Fortey, 1974) in the definition of the rachis and the pleural fur-rows, but differs in having a more semielliptical outline.

The second morphotype (Figs. 29B, D, E, 30B, D, F; and Moberg & Segerberg 1906, Pl. 5:7) is much larger and has a more indistinct transition to the posterior bor-der. The distinct postrachial ridge is absent; only a weakly defined elevation is present postrachially. The Norwegian specimens are found in the grey limestone of the Bjørkåsholmen Formation. The precise level of the Swed-ish specimen is unknown. Morphotype 2 shares the indistinct transition to the border and the ill defined pos-trachial ridge with *Hypermecaspis brevifrons* Fortey, 1974, but differs in having a markedly concave transition to the border region.

Moberg & Segerberg (1906, p. 87) refer to 10 or 12 spec-imens in the collections at Lund University, Sweden,

Table 7. Pygidial measurements of studied specimens of *Tropidopyge broeggeri*.

Specimen	Morph	X	X1	Y	Z	W
LO 1836	1	0.30	–	0.40	0.59	0.91
LO 8371	2	0.70	–	0.94	1.73	2.06
756/1	2	0.32	0.12	0.50	0.87	1.11
69568	1	0.38	0.23	0.51	0.78	0.90
84041/2	2	–	0.24	0.83	1.84	2.40

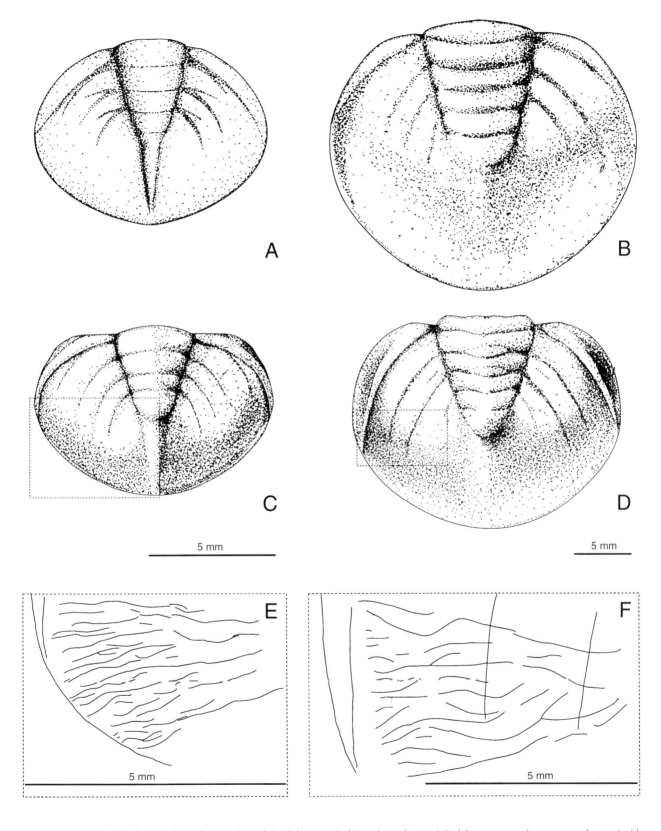

Fig. 30. Reconstructions and comparison of Norwegian and Swedish material of *Tropidopyge broeggeri*. Both have two morphotypes: morphotype 1 with postrachial ridge and morphotype 2, without postrachial ridge. □A. Norwegian specimen (PMO 69568) of morphotype 1. □B. Norwegian specimen (PMO 84041/2) of morphotype 2. □C. Swedish specimen (LO 1836, lectotype) of morphotype 1. Frame indicates area for the exoskeletal detail. □D. Swedish specimen (LO 1837, paralectotype) of morphotype 2. Frame indicates area for the exoskeletal detail. □E. Detail of the terrace ridges of the lectotype. □F. Detail of the terrace ridges of the paralectotype.

ranging from 0.9 to 2.7 cm in length. Only the type specimens from this collection were studied here, since other specimens could not be found (Dr. Per Ahlberg, Lund, written communication, 1994), but a certain morphological variation evidently exists. Combined with the uncertainty of stratigraphical level, the two morphotypes are therefore considered conspecific.

Subfamily Pelturinae Hawle & Corda, 1847

Genus *Peltocare* Henningsmoen, 1957

Type species. – *Acerocare norvegicum* Moberg & Möller, 1898, p. 243, from the upper part of the Alum Shale Formation (upper Tremadoc), formerly Ceratopyge Shale (3aα), at Vækerø near Oslo, Norway; by original designation of Henningsmoen (1957, p. 246).

Discussion. – A total of six species of *Peltocare* is recognized. Henningsmoen (1957, p. 248) regarded *P. glaber* (Harrington, 1938) as conspecific with the type species, while a Mexican form assigned to the type species by Robison & Pantoja-Alor (1968) was assumed to represent a separate species by Nikolaisen & Henningsmoen (1985, p. 22). Hutchinson (1952, p. 94) revised *Cyclognathus rotundifrons* Matthew, 1893, from the Upper Cambrian of eastern Canada, which Henningsmoen (1959, p. 246, 249) assigned to *Peltocare*. Three additional species of this rather uniform genus are known from Norway: the type species *P. norvegicum* from the upper part of the Alum Shale Formation (Tremadoc), formerly Ceratopyge Shale (3aα), *P. modestum* Henningsmoen, 1959, from the overlying Bjørkåsholmen Formation, Oslo Region, and *P. compactum* Nikolaisen & Henningsmoen, 1985, from the lower Tremadoc Berlogaissa Formation, Finnmark, North Norway.

Henningsmoen (1957) distinguished *Peltocare* from the older, closely related *Peltura* by its generally wider fixed cheeks and the anterior facial sutures which cut the cephalic margin further forward. The width (tr.) of the fixed cheeks is, however, not a distinguishing feature. Species like *Peltura costata* (Brøgger, 1882) and *P. transiens* (Brøgger, 1882) have wide (tr.) free cheeks comparable to those of *Peltocare compactum* and *P. modestum*.

The outline of the anterior border is angular in *P. compactum*, weakly so in the younger *P. olenoides* (Salter, 1866) and *P. norvegicum*, and subparallel to the posterior part of glabella in the youngest species *P. modestum*. This may be a transitional feature from young species of *Peltura*, with an angular outline of the anterior border similar to the contemporary *Peltocare compactum*.

Peltocare rotundifrons (Matthew, 1893) appears very similar to *P. modestum*.

Peltocare modestum Henningsmoen, 1959

Figs. 31, 32

Synonymy. – □1959 *Peltocare modestum* n.sp. – Henningsmoen, p. 158, Pl. 1:9, 10. □1988 *Peltocare modestum* Henningsmoen – Rushton, pp. 688, 690, Text-fig. 3A, B.

Holotype. – A cranidium (PMO 69565) from the lower part of the Bjørkåsholmen Formation at Bjørkåsholmen in Asker, Norway. Identified and illustrated by Henningsmoen (1959, p. 158, Pl. 1:9).

Material. – Six cranidia, the largest measuring 0.42 cm (sag. length), and a left free cheek. All specimens were found just above the dark limestone nodules near the base of the formation. Table 8 shows the measurements of available cranidia and Fig. 33 a length/width plot of the cranidia.

Remarks. – Henningsmoen (1959, p. 159) based his diagnosis of the species on three cranidia. Several new specimens, including a free cheek, were made available by Bjørn Funke, Asker. The diagnosis and description have been modified on the basis of the new material.

Emended diagnosis. – A *Peltocare* with narrower fixed cheeks than the type species, more rounded front of glabella, rounded outline of the anterior facial suture and anterior margin. Width (tr.) at palpebral lobes equal to posterior width (tr.) of glabella. Free cheek narrow (tr.) with small pits on distinct border furrow.

Emended description. – Cranidium highly convex (tr.), posterior width nearly 1.5 times sagittal length (Figs. 31A, D, 33). Glabella long, slanting anteriorly with a well-rounded front (Fig. 31B, D). Occipital ring wide (sag.), occupying 20% of glabellar length, with small median node (Fig. 31A, F). The 1S and 2S glabellar furrows are faint, directed obliquely forward. Dorsal furrow well-defined, margin with preglabellar furrow. Posterior part of fixed cheeks twice posterior glabellar width and nearly two-thirds of sagittal length. Posterior border furrow distinct, curving forward laterally. Palpebral lobes small, close to glabella, positioned just posterior of its front. Width (tr.) at eyes equals that of posterior part of glabella. Anterior border narrow (sag. and horizontally), subparal-

Table 8. Cranidial measurements of *Peltocare modestum*.

Specimen	A	B	C	C1	C2	J	J1	J2	K	K1
69565	0.33	0.30	0.05	0.02	0.22	0.53	0.24	0.26	0.25	0.21
69566	0.31	0.26	0.05	0.02	0.20	0.42	0.24	–	0.21	0.18
121.629/3	0.22	0.19	–	–	–	0.29	0.12	–	0.13	0.11
136.080	0.24	0.21	0.03	0.02	0.16	0.33	0.14	0.16	0.16	0.13
136.081/2	0.18	0.20	0.02	0.01	0.26	0.32	0.12	0.13	0.26	0.09
136.081/3	0.42	0.48	0.07	0.05	0.30	0.79	0.30	0.42	0.33	0.21

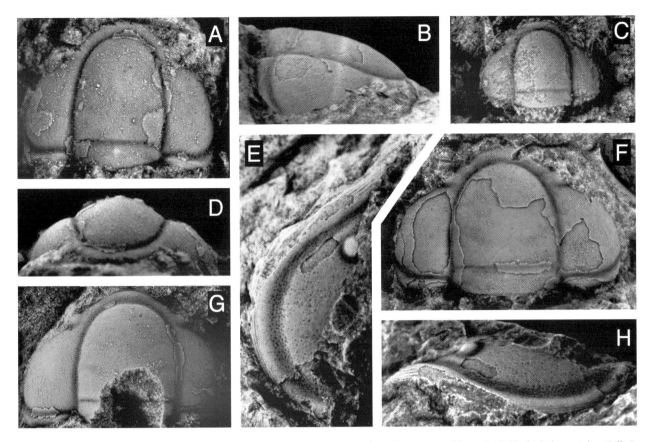

Fig. 31. Peltocare modestum Henningsmoen, 1959. □A, D. Dorsal and anterior view of cranidium; ×18, PMO 121.629/3. Bjørkåsholmen, Asker. Coll.: F. Nikolaisen, 1960-05-01. □B, F. Lateral and dorsal view of large cranidium, showing facial suture; ×8, PMO 136.081/3. Found 30–35 cm above base of the formation at Hagastrand, Asker. Coll.: B. Funke, 1993-04-11. □C. Dorsal view of small cranidium; ×7.5, PMO 136.080. Bjørkåsholmen, Asker. Coll.: B. Funke, 1979-06-02. □E, H. Dorsal and lateral view of free cheek, showing ocular lens and structure of exoskeleton; ×11, PMO 136.081/1. Found 32–35 cm above base of the formation at Hagastrand, Asker. Coll.: B. Funke, 1993-04-11. □G. Dorsal view of holotype cranidium, showing faint glabellar furrows; ×11, PMO 69565. Bjørkåsholmen, Asker. Coll.: G. Henningsmoen, 1958-11-16. Figured by Henningsmoen (1959, Pl. 1:9).

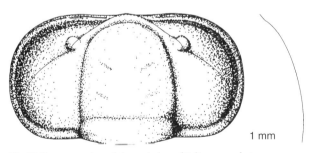

Fig. 32. Reconstruction of the cephalon of *Peltocare modestum*.

1 mm

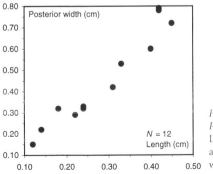

Fig. 33. Cranidia of *Peltocare modestum.* Length (sag.) plotted against posterior width (tr.).

N = 12

Posterior width (cm)

Length (cm)

lel to preglabellar furrow. Anterior branches of facial suture forms the edge of anterior margin. Test smooth (Fig. 31A, F).

Hypostome, thorax and pygidium unknown.

Free cheek typically pelturoid, being elongated, with wide marginal border carrying distinct, densely spaced pits (Fig. 31E). Genal area highly convex (tr.) with anteriorly placed, semispherical compound eyes (Fig. 31E, H). Margin smooth, rounded without spines, carrying marginal terrace ridges.

Fig. 32 shows a reconstruction of the cephalon.

Discussion. – P. modestum is smaller than the type species and differs mainly in having a parabolic outline of the glabella and anterior margin, narrower fixed cheeks (tr.) and anterior border, and a more distinct occipital furrow. It is closer to *P. compactum* Nikolaisen & Henningsmoen, 1985, from the Tremadoc Berlogaissa Formation, Finnmark, Norway, which differs from *P. modestum* in having a less rounded outline of glabella and anterior margin, and a less distinct occipital furrow. These features also distinguish *P. modestum* from *P. olenoides* (Salter, 1866) from the upper Tremadoc of Wales, redescribed by Rushton (1988).

Peltocare rotundifrons (Matthew, 1893) appears very similar to *P. modestum* but has a more angular outline of the anterior border, like *P. compactum*.

As noted by Henningsmoen (1959), the well-rounded anterior margin of *P. modestum* distinguishes it from all other species assigned to *Peltocare*.

Genus *Saltaspis* Harrington & Leanza, 1952

Type species. – *Jujuyaspis steinmanni* Kobayashi, 1936, p. 176, Text-figs. 1–5, from the upper Tremadoc *Triarthrus tetragonalis* – *Shumardia minutula* Biozone of the Saladillo Formation in the Rosario de Lerma department, Salta, Northern Argentina; by original designation.

Discussion. – *Saltaspis* ranges from lower Tremadoc to lower Arenig strata, and includes four species: *S. steinmanni* (Kobayashi, 1936) from the upper Tremadoc of Bolivia and Argentina, *S. vitator* Tjernvik, 1956, from the lower Arenig of Sweden, *S. readingi* Nikolaisen & Henningsmoen, 1985, from the lower Tremadoc of northern Norway, and *S. stenolimbatus* n.sp. from the upper Tremadoc of Norway and Sweden.

Nikolaisen & Henningsmoen (1985, pp. 22–23 for details) argued that *Saltaspis* should be maintained as a separate genus distinguished from *Jujuyaspis* Kobayashi, 1936. This suggestion is followed here. See also Henningsmoen (1957, pp. 244–246) for discussions on the genus and its species.

Saltaspis stenolimbatus n.sp.
Figs. 34, 35

Synonymy. – □1955 *Saltaspis* sp. – Tjernvik, Text-fig. 1D. □1956a *Saltaspis* sp., Tjernvik – Tjernvik, p. 203, Pl. 2:4. □1957 *Saltaspis* sp. Tjernvik – Henningsmoen, pp. 245–246, Pl. 8.

Derivation of name. – From latin *steno* meaning narrow, and *limbatus* meaning bordered, referring to the narrow preglabellar field.

Holotype. – An almost complete cranidium (PMU Vg300), from the Biozone of *Apatokephalus serratus*, Bjørkåsholmen Formation, in profile I at Stenbrottet in Västergötland, Sweden (Fig. 34B, E, F). The specimen was illustrated by Tjernvik (1955, Text-fig. 1D, 1956a, Pl. 2:4).

Paratypes. – A partial cranidium (PMU Vg301) from a loose block at Stenbrottet in Vestergötland (Fig. 34G, H). A cranidium (Fig. 34A, C, D, PMO 144.350) from a dark

limestone nodule near the base of the Bjørkåsholmen Formation at Bygdøy Sjøbad on Bygdøy, Norway.

Other material. – Henningsmoen (1957, p. 246) mentioned a fragment possibly belonging to this species. This specimen has not been located in the collections at the Paleontological Museum in Oslo.

Remarks. – The holotype was collected at level C, 46–52 cm above the base of the Bjørkåsholmen Formation in profile I at Stenbrottet in Västergötland, Sweden. Level C constitutes a shaly unit with small limestone nodules containing *Bienvillia angelini* (Tjernvik 1956a, p. 122). Numerous specimens of *B. angelini* are associated with the holotype; the paratype specimen PMU Vg301 clearly comes from the same stratigraphical level.

Diagnosis. – A *Saltaspis* that differs from the type species in having a broad rectangular glabella, narrow (sag.) preglabellar field and subparallel anterior facial sutures. Palpebral lobe length (sag.) one-third of glabellar length (sag.), posterior margin positioned opposite middle (sag.) of cranidium. Genal spines short(?).

Description. – Sagittal length of cranidium little more than twice the posterior width. Glabella rectangular, broad, flat, slightly bevelled sagittally and transversely before curving relatively steeply down to glabellar furrow (Fig. 34C, E,G). Width (tr.) of glabella two-thirds of length (sag.). Posterior glabellar width (tr.) one-third of posterior width (tr.), slightly more than anterior glabellar width (tr.) in front of palpebral lobes. Posterior lateral part of glabella slightly expanded and rounded. Occipital furrow indistinct, almost transverse medially, branching laterally into indistinct furrows directed obliquely posteriorly and anteriorly, respectively. Posterior branches reaching posterior border. Indistinct median tubercle present. Faint indications of two glabellar furrows (Fig. 34A). Anterior furrow directed obliquely backwards from glabellar furrow just posteriorly of palpebral lobes. Preoccipital furrow positioned laterally opposite middle (sag.) of fixed cheeks. Preglabellar field narrow (sag.), anterior margin subparallel to rounded front of glabella. Anterior facial sutures subparallel. Fixed cheeks near half the length (sag.) and two-thirds the width (tr.) of cranidium. Posterior margin curving gently backwards, genal angle protruding into thin spines of unknown length, but shorter than in other *Saltaspis* species. Terrace ridges present on genal angle. Sutures proparian, subparallel to posterior margin. Posterior border furrow distinct, close to posteriolateral corners of glabella, covering two-thirds (tr.) the length of fixed cheeks before curving slightly anteriorly and vanishing. Reappearing on anterolateral margin of suture. Palpebral lobe length one-third of cranidial length (sag.), posterior margins positioned opposite middle (sag.) of cranidium.

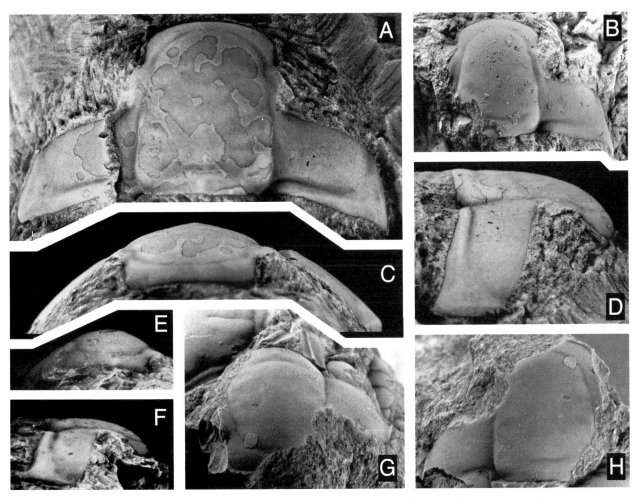

Fig. 34. Saltaspis stenolimbatus n.sp. □A, C, D. Dorsal, anterior and lateral view of paratype; ×5, PMO 144.350. Found in a dark limestone nodule near the base of the formation at Bygdøy Sjøbad, Oslo. Coll.: M. Høyberget, May 1993. □B, F. Dorsal, anterior and lateral view of holotype; ×5, PMU Vg300. Found in profile I at Stenbrottet, Västergötland, Sweden, 46–52 cm above the base. Coll.: T. Tjernvik 1953-08-21. Figured by Tjernvik (1956, Pl. 2:4). □G, H. Anterior oblique and dorsal view of paratype; ×5, PMU Vg301. Found in a loose block at Stenbrottet in Vestergötland, Sweden. Coll.: T. Tjernvik 1953-08-21.

Free cheeks, hypostome, thoracic segments and pygidium unknown.

Discussion. The holotype is the smaller of the three specimens (see Table 9 for measurements). They appear, however, proportionally very similar. In the holotype, the length/width ratio is larger than in the paratype, and the fixed cheeks are not so strongly curved backwards. The faint glabellar depressions that might represent glabellar furrows are only seen in the paratype. Here thin fragments of the exoskeleton are also preserved. It appears to be without structures, except for the terrace ridges on the genal angle. The genal spines diagnostic for the genus are not preserved properly in either specimen. Such spines clearly existed, and though their length is unknown it appears that they might have been shorter than usual for this genus.

The cranidium of *Saltaspis stenolimbatus* n.sp. closely resembles that of the smaller *S. readingi* Nikolaisen & Henningsmoen, 1985, from the lower Tremadoc of northern Norway. The palpebral lobes of this species are proportionally shorter (sag.) and positioned further from the anterior margin. Furthermore, the posterior border furrow curves more strongly obliquely forward and the genal spines are more prominent. This is also the case for the type species, *S. steinmanni* (Kobayashi, 1936) from the upper Tremadoc of Argentina. The type species differs, however, also in having a proportionally narrower glabella with a more rounded front. The preglabellar field does not conform to the shape of the glabella and the preglabellar field is proportionally wider (sag.). The fixed cheeks are also straighter and the palpebral lobes positioned further away from the glabella. In *S. vivator*

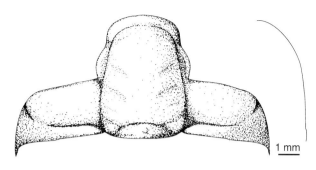

Fig. 35. Reconstruction of the cranidium of *Saltaspis stenolimbatus* n.sp.

Table 9. Cranidial measurements of *Saltaspis stenolimbatus* n.sp.

Specimen	A	B	C	C1	C2	J	J1	J2	K	K1
PMU Vg 300	0.61	0.57	0.15	0.15	0.29	1.10	0.26	0.35	0.46	0.44
PMU Vg 301	~0.75	0.66	–	–	–	1.36	0.50	–	0.58	–
144.350	0.88	0.85	0.24	0.24	0.37	1.90	0.79	0.59	0.68	0.66

(Tjernvik, 1956) from the Early Arenig of Sweden, the palpebral lobes are positioned opposite the front of the glabella and the fixed cheeks are proportionally wider (sag.). This is the only other species with indications of glabellar furrows.

Fig. 35 presents a reconstruction of *Saltaspis stenolimbatus* n.sp.

Subfamily Oleninae Burmeister, 1843

Genus *Bienvillia* Clark, 1924

Type species. – *Dikelocephalus? corax* Billings, 1865, p. 334, Fig. 322, from an Upper Cambrian limestone boulder in the Levis Formation at Levis in Quebec, Canada; by original designation of Clark (1924).

Discussion. – Specimens described below are now assigned to *Bienvillia* rather than to *Triarthrus*. Some forty species have been ascribed to *Triarthrus* from Tremadoc to Ashgill strata, but many of these have been reassigned to other genera. The genus is famous because of the pyritized specimen of *T. eatoni* (Hall, 1838), from the Utica Shale near Rome, New York, showing the appendages (Whittington & Almond 1987, and references therein). Fortey (1974, p. 70) expressed the need for a revision of many of the known species.

Ludvigsen & Tufnell (1983) found that the taxonomic base of *Triarthrus*, concerning its generic longevity, was not well founded. They reassigned species with no preglabellar field and an anterior border furrow to the genus *Porterfieldia* Cooper, 1953, and the species with a preglabellar field and an anterior border furrow to the genus

Bienvillia Clark, 1924. These features were also discussed by Fortey (1974, p. 65). *Triarthrus* lacks an anterior border furrow on the cranidium. These characters were not, however, used in separating *Triarthrus* from closely related olenids, and the three above-mentioned genera were considered a monophyletic group, with *Bienvillia* as the ancestral form.

The revision of the genus *Triarthrus* by Ludvigsen & Tufnell (1983) has led to a change in nomenclature of species regarded as typical of *Triarthrus*, such as *T. angelini* Linnarsson, 1869, and *T. parchaensis* Harrington & Leanza, 1957, both assigned to the genus *Bienvillia*.

Bienvillia angelini (Linnarsson, 1869)

Figs. 36, 37

Synonymy. – □1869 *Triarthrus angelini* n.sp. – Linnarsson, p. 70, Pl. 2:28. □1882 *Triarthrus angelini* Linnarsson – Brøgger, p. 112, Pls. 3:1, 1a; 12:1, 1a. □1906 *Triarthrus angelini* Linnarsson – Moberg & Segerberg, p. 83, Pl. 4:29–31. □1957 *Triarthrus angelini* Linnarsson – Harrington & Leanza, p. 115. □1957 *Triarthrus angelini angelini* Linnarsson – Henningsmoen, pp. 148–149, Pls. 8; 11:8–10. □1974 *Triarthrus angelini* Linnarsson – Fortey, p. 65. □1983 *Bienvillia angelini* (Linnarsson) – Ludvigsen & Tufnell, p. 574.

Type material. – The cranidium figured by Linnarsson (1869, Pl. 2:28), locality not mentioned, is probably lost. A neotype is at present not selected from the Swedish material at the Museum of Evolution, Uppsala, and the Geological Survey of Sweden, Uppsala. Until such times as the type is found or a neotype is selected, the present material must suffice to define the taxon.

Norwegian material. – Cranidia are commonly encountered in the dark limestone nodules near the base of the Bjørkåsholmen Formation across the Oslo Region, but also some free cheeks, pygidia and thoracic segments are found. A few specimens are also found ranging throughout the unit. The largest cranidium measures 0.95 cm (sag. length).

Diagnosis. – See Henningsmoen (1957, pp. 148–149).

Description. – The species has never been properly described. A full description is therefore provided here based on the present material.

Sagittal length of cranidium three-fourths of posterior width. Glabella long and wide, expanding slightly forward, making up 90% of sagittal length and two-thirds of posterior width. Occipital ring well defined, with small median node. Central part of occipital furrow crescentic, lateral parts curving obliquely forward. The 1S and 2S furrows are parallel, lateral parts curving anteriorly, adrachial parts curving posteriorly from dorsal furrow at

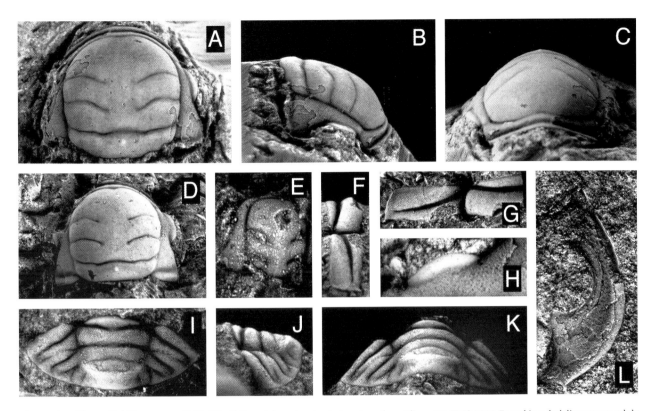

Fig. 36. Bienvillia angelini (Linnarsson, 1869). □A–C. Dorsal, lateral and anterior view of cranidium; ×5, PMO 1244. Found in a dark limestone nodule 25–30 cm above base of the formation at Bjørkåsholmen, Asker. Coll. unknown, 1915. □D. Dorsal view of cranidium, showing glabellar furrows; ×6, PMO 66831/1. Found in a dark limestone nodule 5–10 cm above base of the formation at Skarra in Vestfossen, Øvre Eiker. Coll.: G. Henningsmoen, 1954-04-07. Figured by Henningsmoen (1957, Pl. 11:10). □E. Dorsal view of meraspid stage; ×25, PMO 121.629/4. Found in a dark limestone nodule 25–30 cm above base of the formation at Bjørkåsholmen, Asker. Coll.: F. Nikolaisen, 1960-05-01. □F, G. Lateral and dorsal view of left half of a thoracic segment; ×6, PMO S1277. Found in a dark limestone nodule 32–37 cm above base of the formation at Vækerø, Oslo. Coll.: L. Størmer, 1919? □H. Dorsal view of palpebral lobe; ×10, PMO 84052. Found in a dark limestone nodule 5–10 cm above base of the formation at Skarra in Vestfossen, Øvre Eiker. Coll.: G. Henningsmoen, 1961-10-24. □I, J, K. Dorsal, lateral and posterior view of pygidium; ×11, PMO 66830. Found in a dark limestone nodule 5–10 cm above base of the formation at Skarra in Vestfossen, Øvre Eiker. Coll.: G. Henningsmoen, 1954-04-07. Figured by Henningsmoen (1957, Pl. 11:8). □L. Dorsal view of right free cheek showing the small posteriolateral spine; ×9.5, PMO 144.353. Found in a dark limestone nodule near the base of the formation at Bygdøy Sjøbad, Oslo. Coll.: M. Høyberget, May 1993.

145°, not meeting across glabella. Pair of small pits present anteriorly of adrachial end of S2 furrows, midway between 2S furrows and preglabellar furrow (Fig. 36D). Anterior glabellar lobe well rounded, curving slightly backwards sagittally. Dorsal and preglabellar furrows thin, but distinct. Posterior part of fixed cheeks narrow (tr.), tapering towards palpebral lobes, making up two-fifths of sagittal length. Posterior margin curving forward adrachially. Posterior border furrow transverse, distinct. Palpebral lobes semielliptical, elongated, making up nearly one-third of sagittal length, centred midway between 1S and 2S furrows. Preglabellar area, anterior border furrow and anterior border narrow (tr.) subparallel to anterior part of glabella. Exoskeleton on fixed cheek finely granulated (Fig. 36H).

Free cheek narrow (tr.), elongated. Border furrow vague. Border with marginal terrace ridges. Posterior bor-

der curves forward to short genal spine pointed outwards at a right angle to the border (Fig. 36L).

Thoracic segments with narrow pleural fields. Pleural furrows distinct, curving obliquely backwards, attenuated laterally.

Sagittal length of pygidium two-fifths of maximum width (tr.). Rachis almost as wide as long, tapering slightly posteriorly. It carries three rachial rings, the posterior one almost fused with bluntly rounded posterior rachial lobe (Fig. 36J). Pleural fields sloping steeply laterally, carrying three pleurae. These are directed obliquely backwards parallel to articulating facets. Interpleural furrows weekly indicated (Fig. 36I). Margin semielliptical, changing direction and converging strongly backwards at maximum width, being opposite second rachial furrow. Close to rachis posterior margin short, articulating facets occupying lateral parts. Faint terrace ridges present at postero-

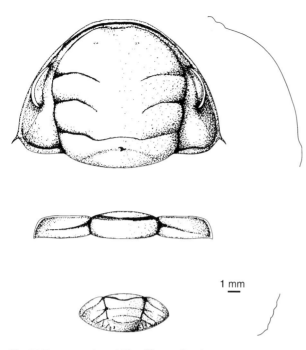

Fig. 37. Reconstruction of *Bienvillia angelini* showing cephalon, a thoracic segment and the pygidium.

lateral corners of margin. Exoskeleton covered with fine granules (Fig. 36I).

Ontogeny. – The meraspid cranidium of *Bienvillia* is similar to the holaspid stage in the general outline but differs markedly in the characters of the glabella (Fig. 36E). It is subquadratic in outline; the anterior glabellar lobe is more bluntly rounded, and the glabellar furrows are deeper and straight. The anterolateral area of the fixed cheeks is much wider (tr.).

The posterior rachial ring of the pygidium does not appear until very late in the ontogeny, and the interpleural furrows become gradually more distinct (Fig. 36I, J).

Discussion. – Following Ludvigsen & Tufnell (1983), this species is assigned to *Bienvillia* rather than to *Triarthrus*. *T. angelini rectifrons* Harrington, 1938, from Argentina was upgraded to a specific level by Harrington & Leanza (1957), thus eliminating the need of the subspecific distinction of *T. angelini angelini sensu* Henningsmoen (1957).

B. angelini is very similar to coeval species like *B. tetragonalis* (Harrington, 1938), from the Tremadoc of Argentina, and *B. shinetonensis* (Lake, 1913), from the Tremadoc of Britain and Argentina, differing from these generally in having a more rounded anterior part of glabella and a narrower anterior part of the fixed cheeks. These early species of *Bienvillia* are rather uniform, distinguished mainly by minor morphological details.

B. angelini is a diagnostic element within the dark limestone nodules near the base of the Bjørkåsholmen Formation all over the Oslo Region. Fig. 37 presents a reconstruction of *B. angelini*.

Genus *Parabolinella* Brøgger, 1882

Type species. – *Parabolinella limitis* Brøgger, 1882, p. 103, Pl. 3:2, from the uppermost part of the Alum Shale Formation (lower Tremadoc), formerly Ceratopyge Shale (3aβ), at St. Olavsgate in Oslo, Norway; subsequently designated by Bassler (1915, p. 943).

Discussion. – A diagnosis of the genus was presented by Robison & Pantoja-Alor (1968) and emended by Ludvigsen (1982). Rushton (1988) discussed morphological characters from which *Parabolinella* could be further recognized, and listed the species described since the revision of this genus by Henningsmoen (1957). He pointed out (Rushton 1988, p. 686) that several Chinese species described are likely to be synonyms.

Parabolinella Brøgger, 1882, *Parabolina* Salter, 1849, and *Bienvillia* Clark, 1924 are closely related. The phylogeny and relationship of these genera were discussed in detail by Henningsmoen (1957, p. 134) and are not repeated here.

The exact stratigraphical range of the predominantly Tremadoc genus *Parabolinella* is not known, but if the species *P.? caesa* Lake, 1913 (see also Morris & Fortey 1985, p. 108, Pl. 2:2) and *P.? simplex* (Salter, 1866), both Upper Cambrian species from North Wales, should prove to belong in this genus, then the coeval genus *Parabolina* is not the ancestral form suggested by Henningsmoen (1957, p. 135).

Parabolinella lata Henningsmoen, 1957
Fig. 38

Synonymy. – □?1956a *Parabolinella* sp. – Tjernvik, p. 200. □1957 *Parabolinella lata* n.sp. – Henningsmoen, p. 135, Pls. 8; 12:8. □non 1984 *Parabolinella lata* Henningsmoen – Xiang & Zhang, p. 402, Pl. 1:9, 10. □1988 *Parabolinella lata* Henningsmoen – Rushton, pp. 686, 687.

Holotype. – A small, slightly damaged cranidium (PMO 1287a) from the Bjørkåsholmen Formation at Bjørkåsholmen, Asker, Norway. Identified and figured by Henningsmoen (1959, p. 135, Pls. 8; 12:8).

Material. – Five cranidia. See Table 10 for measurements.

Diagnosis. – See Henningsmoen (1957, p. 135).

Emended description. – The description given by Henningsmoen (1957) is adequate for the single specimen he

Fig. 38. Parabolinella lata Henningsmoen, 1957. □A, C. Dorsal and anterior view of holotype cranidium showing eye ridges and anterior margin; ×10, PMO 1287/1. Bjørkåsholmen, Asker. Coll. unknown, 1915. Figured by Henningsmoen (1957, Pl. 12:8). □B, D. Dorsal and lateral view of incomplete cranidium, showing glabellar furrows and anterior border furrow pits (internal mould); ×4, PMO 115.322/1. Bjørkåsholmen, Asker. Coll.: G. Henningsmoen, 1958-11-23. □E. Dorsal view of incomplete cranidium, showing anterior area, border and anterior glabellar furrow; ×5.5, PMO 97272/2. Slemmestadveien 6 in Slemmestad, Røyken. Coll.: M. Havrevoll & J. Gjessing, 1975. □F. Dorsal view of latex replica from external mould of incomplete cranidium, showing fixed cheek; ×3, PMO 121.632. Bjørkåsholmen?, Asker. Coll.: J. Gjessing, 1976. □G. Dorsal view of latex replica from external mould of incomplete cranidium, showing fixed cheek and glabellar furrows; ×2, PMO 84051. Dark limestone nodule 5–10 cm above base of the formation at Skarra in Vestfossen, Øvre Eiker. Coll.: G. Henningsmoen, 1961-10-24.

assigned to *Parabolinella lata*. The other cranidia assigned here to *P. lata* are all larger and justify a revision of the description.

Posterior width of cranidium twice sagittal length. Glabella subrectangular, slightly wider (tr.) than long (sag.), making up nearly two-thirds of sagittal length and one-third of posterior width (Fig. 39). Median occipital node

faintly indicated in large specimens (Fig. 38G). Occipital furrow well defined, central part transverse, lateral parts directed obliquely forward, not reaching dorsal furrow. The 1S furrows are distinct, converging rearwards at an angle of 155°, bifurcating laterally. The 2S furrows are directed obliquely backwards, subparallel to 1S furrows, slightly sinuous. The 3S furrows are short, transverse, sit-

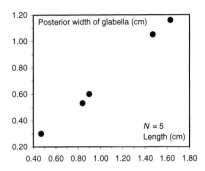

Fig. 39. Cranidia of *Parabolinella lata.* Length (sag.) plotted against posterior width (tr.) of glabella.

Table 10. Cranidial measurements of *Parabolinella lata.*

Specimen	A	A1	B	C	C1	C2	J	J1	J2	K	K1
1287/1	0.47	0.45	0.33	0.12	0.08	0.23	–	0.6	0.63	0.3	0.32
83796	–	–	–	–	–	–	–	–	–	–	2.42
84051	1.63	–	1.38	0.44	0.02	0.85	3.24	1.78	2.20	1.16	1.23
97272/2	0.84	0.80	0.69	–	–	–	–	–	–	0.53	0.56
115.322/1	0.90	0.83	0.65	–	–	–	–	–	–	0.60	0.65
121.632	1.47	–	1.07	0.45	0.24	0.78	2.97	1.35	1.74	1.05	0.81

uated midway between 2S furrows and preglabellar furrow. The 4S furrows are small, pit-like, situated laterally at anterolateral corners of glabella. Anterior glabellar lobe curving slightly backwards sagittally (Fig. 38B, E). Posterior part of fixed cheek wide (tr.), horizontal adrachially, curving steeply downwards laterally. The cheek makes up two-fifths of sagittal length. Posterior border furrow and border well defined, posterior margin diverging slightly obliquely backwards. Width at palpebral lobes slightly more than half posterior width. Palpebral lobes small, semicircular, merging with short, slightly obliquely posterior directed eye ridges. Situated a distance from glabella, ranging from 3S furrows to midway between 2S furrows and anterior branch of 1S furrows. Anterior branches of facial suture diverging anteriorly. Preglabellar area convex, curving steeply down to anterior border furrow with many pits. Anterior border and border furrow semielliptical, defining a less steeply curved terrace. Reticulated pattern on preglabellar exoskeleton (Fig. 38B, E). Test on glabella shows faint fingerprint-like pattern.

Discussion. – The smallest cranidium of *P. lata* is 0.46 cm long (sag.) and is regarded as a juvenile specimen, probably meraspid stage (Fig. 38A, C). The other cranidia assigned here are 0.67 cm to more than 1.5 cm long (sag.) (Fig. 38E and G, respectively). The smaller cranidium differs in having a larger median occipital node, a transverse occipital furrow, no S3 and S4 furrows, and palpebral lobe positioned more anteriorly, just opposite S2 furrow. Most of these differences are related to ontogeny. The relative size of the median occipital node is usually reduced, and also the position and size of the palpebral lobes can be altered (Harrington *et al. in* Moore 1959, pp. O141, O139). The 3S and 4S furrows are more distinct in the

larger specimens (Fig. 38G), compared with the faint indications of these furrows on smaller specimens (Fig. 38E). The larger specimens assigned to *P. lata* resemble the type species, *P. limitis,* except that the median occipital node is fainter, the palpebral lobe is shorter and positioned more anteriorly, ranging from S3 furrow to midway between S2 and anterior branch of S1 furrows, the anterior border furrow has more distinct pits, and the preglabellar area shows a reticulated exoskeletal pattern.

P. lata and *P. limitis* are evidently very closely related, but their stratigraphical positions and the morphological differences are here regarded as sufficient to warrant a separation. Henningsmoen (1957, p. 133) also emphasized the very close relationship between the Tremadoc species *P. argentinensis* Harrington & Leanza, 1957, and *P. triarthroides* Harrington, 1938, from Argentina and North Wales, *P. triarthrus* (Callaway, 1877), from the Welsh Borderland, and *P. limitis,* from Scandinavia, and suggested that *P. lata* was a descendant from this group.

Xiang & Zhang (1984) described *P. lata* n.sp. from the Xinjiang Province of China. However, this species differs markedly from the Norwegian material assigned to *P. lata* in having an occipital furrow and glabellar furrows that reach the dorsal furrow, no bifurcating of 1S furrows, small palpebral lobes situated between 2S and 3S furrows and a straight posterior part of the facial suture.

Family Remopleurididae Hawle & Corda, 1847

Subfamily Richardsonellinae Raymond, 1924

Genus *Apatokephalus* Brøgger, 1896

Type species. – *Trilobites serratus* Boeck, 1838, pp. 139–140, from the Bjørkåsholmen Formation (upper Tremadoc) near the old Aker Church in Oslo, Norway. Subsequently designated by Bassler (1915, p. 55).

Remarks. – Despite being a type species, *Apatokephalus serratus* has been rather poorly understood. Several forms have been erroneously attributed to this species (see discussion below for further details). Tjernvik (1956a) and Harrington *et al. in* Moore (1959) provided general diagnostic characters for *A. serratus.* However, a more adequate diagnosis of this species is given here, based on the well-preserved Norwegian material.

Emended diagnosis. – 1S furrows close to margin of glabella, transverse laterally, geniculating backwards adrachially. Preglabellar area narrow (sag.), with wide (sag.) anterior border tapering laterally into narrow fixed cheeks (sag.). Smooth transition to anterior border furrow. Glabella covered with prominent, large and openly spaced tubercles, filling the space between the palpebral lobes. Pygidium with pygidial spines arranged marginally in a gentle posteriorly directed arcuate fashion. Maximum width (tr.) opposite posterior part of rachis. Pleural fields prominently step-like.

Discussion. – Chugaeva (*in* Chugaeva *et al.* 1973, p. 49) provided a rather general diagnosis of this large cosmopolitan genus, including both species with and without a preglabellar area. Peng (1990, p. 92) discussed the current concept of the genus and noted variations in the width (sag.) of the preglabellar field and in the nature of the glabellar furrows. To this can be added a variation in the sculpture of the glabella, ranging from granules of varying size and numbers (*A. serratus*), to fine, raised lines of different patterns (*A. dubius*), and the direction of the posterior facial sutures ranging from gently diverging curves (*A. exiguus*), to straight, almost backward-curving sutures (*A. serratus*). However, the anterior suture always outline distinct laterally tapering anterior areas of fixed cheeks with a rounded anterior margin, and usually cuts the margin close to (exsag.) or opposite the front of the anterior of glabella.

Several genera similar to *Apatokephalus*, for instance *Elkanaspis*, *Pseudokainella*, *Apatokephaloides*, and *Apatokephalops*, are placed in separate families, but it it is easy to appreciate overall morphologial similarities. Ludvigsen (1982) erected the North American genus *Elkanaspis* of the Family Kainellidae Ulrich & Resser, 1930, which bears some similarity to *Apatokephalus*, especially in the expression of the glabellar features and the general look of the pygidium. Mature specimens of the latter genus lack the prominent intraocular parts of the fixed cheeks, generally isolated by the glabellar and palpebral furrows. *Elkanaspis* also lacks the distinct laterally tapering anterior fixed cheeks seen in *Apatokephalus*; instead it has diverging anterior sutures cutting the anterior margin anterior to the front of the glabella. The prominent steplike pleural fields of the pygidium are not so well developed as in *Apatokephalus*. However, the British species *Apatokephalus sarculum* has distinct interocular fixed cheeks in the early ontogeny (Fortey & Owens 1991, p. 455, Fig. 12a, f), and an overall morphology similar to that of *A. serratus*. Thus, this fact alone emphasizes the close relationship between *Apatokephalus* and *Elkanaspis*, and for that matter other genera similar to these two. The question of generic similarities also leads to the question of family affinities, and the phylogeny and relationship of *Apatokephalus* and related genera must clearly be revised in a broader context.

However, *Apatokephalus* has numerous species, most of which are poorly known or poorly related to the type species, which itself has been less than satisfactory described. An attempt is made here to present the current concept of *Apatokephalus*, but without addressing intrageneric relationships.

Species belonging to this genus were listed by Peng (1990, p. 91). Additional species are *A. multispinosus* (Raymond, 1937) from the Tremadoc of Vermont, USA; *A. gillulyi* Ross, 1958, from the lower Tremadoc of Nevada, USA; *A. serratus serratus sensu* Poletaeva *in* Petrunina *et al.* (1960) from the Tremadoc of Sayan Altai, Russia; *A. serratus dubius sensu* Balashova (1961) from the Tremadoc of the Aktyubinsk Province, Russia; *A. globosus* Chugaeva *in* Chugaeva *et al.* 1964; *A. incisus* Dean, 1966, from the Lower Ordovician of Montagne Noir, France; *A. wilsoni* Dean, 1966, from the Lower Ordovician of Texas, USA; *A. serratus pamiricus* Balashova, 1966, from the Lower Ordovician of Russia; *A. heterosulcatus* Burskij, 1970, from the Lower Ordovician of Pai–Khoj, Russia; *A. serratus*, *A.* sp. 1 & sp. 2 *sensu* Burskij (*in* Bondarev *et al.* 1965; 1966, 1970) from the lower Tremadoc of Pai–Khoj, Russia; *A.* sp *sensu* Ross (1970) from the lower Middle Ordovician of Nevada, USA; *A.* sp. *sensu* Dean (1973) from the Arenig of the Taurus Mountains, Turkey; *A. tianshifuensis* Kou *in* Kou *et al.* 1982, from the Tremadoc of China; *A. dagmarae* Mergl, 1984, from the upper Tremadoc of Central Bohemia, Czech Republic; *A. hunjiangensis* Duan & An, 1986, from the upper Tremadoc, southern Jilin, China; *A.*? *longifrons* Dean, 1989, from the lower Tremadoc, Alberta, Canada; *A. latilimbatus* Peng, 1990, from the lower part of the Madaoyu Formation (upper Tremadoc), Hunan, S. China; *A. sarculum* Fortey & Owens, 1991, from the upper Tremadoc of the Welsh Borderland, *A. fedoui* Pillet, 1992, from the Lower Ordovician of Montagne Noir, France, *A.*? sp. *sensu* Mergl (1994) from the Tremadoc of Central Bohemia, Cezch Republic, and *A. dactylotypos* n.sp. and *A.* cf. *sarculum* from the upper Tremadoc of Norway (see Fig. 40 for distribution).

The following species are assigned to other genera:

A. assai Weber, 1932, and *A.* sp. 1 (Chugaeva *in* Chugaeva *et al.* 1973) belong to *Eorobergia* Cooper, 1953 (Nikolaisen 1991, p. 51). Likewise, *Apatokephalus*? *octopoides* Kobayashi, 1934 and *A.* sp. *sensu* Dean (1989) (Canada) are here assigned to this genus; *Apatokephalus sibericus* Rozova, 1960 (Siberia), and *A. nyaicus* Rozova, 1968 (Siberia), were assigned to *Eoapatokephalus* Rozova, 1983, by Rozova (1983). Peng (1990, p. 91) assigned *Apatokephalus sensu* Reed (1903) to the genus *Robergia* Wiman, 1905, *Apatokephalus concavemarginatus* Lu *in* Zhou *et al.* 1977, to the genus *Apatokephalops* Lu, 1975, and *Apatokephalus jiangshanensis* Lu & Lin, 1984, to the genus *Remopleuridiella* Ross, 1951. *Apatokephalus* sp. *sensu* Chang & Fan (1960) may also be referred to the latter genus. Pillet (1992) made *Apatokephalus brevifrons* (Thoral, 1935) the type species of the genus *Dekanokephalus* Pillet, 1992.

		GONDWANALAND Asia, Europe, South America	AVALONIA Wales	LAURENTIA North America	BALTICA Scandinavia	SIBERIA Russia & Kahzakstan
ORDOVICIAN	ARENIG	A. sp. *sensu* Dean (1973) A. incisus Dean, 1966			A. pecten Wiman, 1905	
ORDOVICIAN	TREMADOC	A. fedoui Pillet, 1992 A. cf. tibicen Přibyl & Vaněk, 1980 A. tibicen Přibyl & Vaněk, 1980 A.? sp. sensu Mergl (1994) A. latilimbatus Peng, 1990 A. hunjiangensis Duan & An, 1989 A. dagmarae Mergl, 1984 A. tianshifuensis Kou, 1982 A. invitatus Liu in Zhou et al., 1977 A. bellus Liu in Zhou et al., 1977 A. yini Lu in Chang & Fan, 1960 A. kansuensia Chang & Fan, 1960 A. asarkus Sduzy, 1955 A. hyotan Kobayashi, 1953 A. exiguus Harrington & Leanza, 1957	A. sarculum Fortey & Owens, 1991 A. sp. sensu Rushton (1980)	A. sp. sensu Ross (1970) A. gilluly Ross, 1958 A.? longifrons Dean, 1989 A. finalis (Walcott, 1884) A. wilsoni Dean, 1966 A. canadensis Kobayashi, 1953 A. levisensis (Rasetti, 1943) A. multispinosus (Raymond, 1937)	A. dubius (Linnarsson, 1869) A. serratus (Boeck, 1838) A. dactylotypos n. sp. A. cf. sarculum Fortey & Owens, 1991: Ebbestad	A. sp. sensu Rosova (1985) A. serratus Boeck sensu Burskij (1970) A. serratus pamiricus Balashova, 1966 A. globosus Chugaeva, 1964 A. poletavea Fedyanina, 1960 in Petrunina et al. (1960) A. schoriensis Poletaeva, 1960 in Petrunina et al. (1960) A. serratus serratus sensu Poletavea, 1960 in Petrunina et al. (1960) A. replicare Lisgor, 1954 A. hetrosulcatus Burskij, 1970 A. sp. 1 sensu Burskij (1970) A. sp. 2 sensu Burskij (1970) A. serratu serratus sensu Balashova (1961)

Fig. 40. Distribution of species attributed to *Apatokephalus.*

A. serratus fedoui Pillet, 1992 (France) is here regarded a distinct species and not a subspecies of *A. serratus* (Boeck, 1838). Pillet (1992) stated that only subtle details separated the two species; however, the French species is distinguished by having obliquely forward pointed posterior glabellar furrows, deeper (tr.) middle glabellar furrows, straighter and deeper (tr.) preoccipital glabellar furrows, no preglabellar area, and a bevelled anterior border.

In an attempt to untangle the problems with the great number and varieties of *Apatokephalus* species, Pillet (1992) subdivided the genus in two groups of subgeneric status; *Apatokephalus* (*Apatokephalus*) and *A.* (*Tibikephalus*). The first group included species with a general recemblance to the type species. The latter group was centred around *A. tibicens* Přibyl & Vaněk, 1980, and comprised some of the characteristic American species attributed to *Apatokephalus* (see Pillet 1992, pp. 24–25, for lists of included species). The subdivision was based entirely on cranidial characters, pygidial characters being disregarded because pygidia are unknown for a number of *Apatokephalus* species (Pillet 1992, p. 24). The diagnostic cranidial characters of *A.* (*Tibikephalus*) included a less expanding frontal part of glabella and a less sharp diverging course of the posterior facial sutures. However, within the morphological variation of the genus intermediate

forms are easy to find, such as *A. yini* Lu *in* Chang & Fan, 1960, and *A. sarculum* Fortey & Owens, 1991, with a gentle diverging curve of the facial sutures and a moderately expanded front of the glabella. Both species were included in *A.* (*Apatokephalus*) by Pillet (1992). The expression of the frontal glabellar lobe, the facial sutures and the glabellar furrows are highly variable within the concept of *Apatokephalus*, as also noted by Peng (1990). The concept of subgenera within *Apatokephalus* adapted by Pillet (1992) is therefore considered ambiguous, even more so in the light of the pygidial characters, and is better not applied.

It is, however, agreed that two relatively distinct groups exist within *Apatokephalus*. This assumption is based on somewhat other reasons than those presented by Pillet (1992) and includes the pygidial characters which are here considered important for a separation. A possible configuration of the genus *Apatokephalus* is outlined below. This is only meant as a general discussion, not a proposal for a formal separation, for which the data are inadequate.

Some species referred to *Apatokephalus* are represented by poorly preserved or fragmentary material from which it is difficult to distinguish separating characters, e.g., *A. serratus pamiricus* Balashova, 1966 (Siberia), *A. serratus dubius* Balashova (1960) (Siberia), *A. sp.* 1 Burskij (1970) (Siberia), *A. schoriensis* Poletaeva *in* Petrunina *et al.*, 1960

(Siberia), *A. poletaevae* Fedyanina *in* Petrunina *et al.*, 1960 (Siberia), and *A.* sp. *sensu* Dean (1973).

The group of trilobites centred around *A. serratus* are generally recognized by having distinct glabellar furrows where the anterior glabellar furrow is almost straight. The pygidia of species within this group have four to six pairs of marginal spines becoming progressively shorter towards the centre. Otherwise the variation is great for pygidial size, rachial size, expression of pleural furrows, ornament, and so on. The group is believed to comprise the following species:

A. finalis (Walcott, 1884); *A. dubius* (Linnarsson, 1869) (Norway and Sweden); *A. levisensis* (Rasetti, 1943) (Canada); *A. canadensis* Kobayashi, 1953 (Canada); *A. replicare* Lisogor, 1954 (Kazakhstan); *A. asarkus* Sdzuy, 1955 (Bohemia); *A. exiguus* Harrington & Leanza, 1957 (Argentina); *A. wilsoni* Dean, 1966 (North America); *A. serratus sensu* Burskij (*in* Bondarev *et al.* 1965; 1966; 1970), *A. tibicen* Přibyl & Vaněk, 1980 (Argentina); *A.* sp. Rushton (1982) (North Wales); *A. dagmarae* Mergl, 1984 (Bohemia); *A. longifrons* Dean, 1989, *A. latilimbatus* Peng, 1990 (South China); *A. sarculum* Fortey & Owens, 1991 (North England); *A. dactylotypos* n.sp. (Norway) and *A.* cf. *sarculum* (Norway).

A typical representative of the second group is *A. heterosulcatus* Burskij, 1970 (Siberia).

Compared with *A. serratus*, these trilobites have an anterior glabellar furrow which is either indistinct or clearly directed obliquely backwards. Where the pygidium is known, the marginal pleural spines inside the second or third pair are many and short, usually arranged in an anteriorly directed arcuate fashion.

The group is believed to comprise the following species: *A. pecten* Wiman, 1905 (Sweden, ?Siberia); *A. multispinosus* (Raymond, 1937); *A.? hyotan* Kobayashi, 1953 (Korea); *A. kansuensis* Chang & Fan, 1960 (China); *A. yini* Lu *in* Chang & Fan, 1960 (China); *A. gillulyi* Ross, 1958 (USA), *A. globosus* Chugaeva *in* Chugaeva *et al.* 1964 (cranidium only) (northeastern former USSR); *A. incisus* Dean, 1966 (France); *A.* sp. *sensu* Ross (1970) (USA); *A.* sp. 2 *sensu* Burskij (1970, pygidium only) (Siberia), *A. hunjiangensis* Duan & An, 1986 (North China) and *A. fedoui* Pillet, 1992 (France).

The separation made above is, however, not straightforward. In species like *A.* sp. *sensu* Ross (1970), the cranidium seems to fit best into the second group. The pygidium has shorter marginal spines near the centre, but they are not numerous or arranged in a crescentic curve. The same applies for *A.? hyotan* Kobayashi 1953. The cranidium conforms to the second group, while the pygidium is somewhere between the groups having only three short marginal spines near the centre. This is also the case for *A. incisus* Dean, 1966. The pygidium of *A. sarculum* Fortey & Owens, 1991, conforms to the first group,

but has also some resemblance to the pygidium of *A.* sp. *sensu* Ross (1970).

Nikolaisen (1991, p. 51) noted the close similarity of *A. hunjiangensis* to *Eorobergia breviceps*. He believed that the younger *E. marginalis* group of trilobites has developed from *Apatokephalus* or closely related species. Typical species of the second group, such as *A. pecten* and *A. globosus*, are also quite similar to the type species of *Eorobergia*, *E. marginalis*. It may therefore seem that the second group of trilobites have a closer affinity to *Eorobergia*, distinct from the species of *Apatokephalus* centred around the type species. However, considering the wide range of morphotypes assigned to both genera, a separation is difficult.

The genus *Apatokephalus* is in need of a comparative treatment of the whole group. The concept of Pillet (1992), however, does not suffice, and a further development of the scheme presented here may prove more fruitful.

Fig. 41 presents reconstructions of the type species and other species of *Apatokephalus* known from the Lower Ordovician Bjørkåsholmen and Alum Shale formations. The species are described and discussed in the following paragraphs.

Apatokephalus serratus (Boeck, 1838)

Figs. 41A, 42

Synonymy. – ☐1838 *Trilobites serratus* n.sp. – Boeck, p. 139. ☐1854 *Centropleura serrata* (Boeck) – Angelin, p. 88, Pl. 41:10. ☐1854 *Centropleura angusticauda* n.sp. – Angelin, p. 88, Pl. 41:10*. ☐1869 *Dikelocephalus angusticauda* (Angelin) – Linnarsson, p. 71. ☐1882 *Dicelocephalus serratus* (Boeck) – Brøgger, p. 126, Pl. 3:7, 8. ☐1896 *Apatocephalus serratus* (Boeck) – Brøgger, pp. 184–185. ☐1896 *Apatokephalus dubius* (Linnarsson) – p. 175, Fig. 5a, b. ☐1897 *Dicellocephalus serratus* (Boeck) – Holm, p. 465, Pl. 8:3–5. ☐1906 *Apatocephalus serratus* (Boeck) – Moberg & Segerberg, pp. 88–89, Pl. 5:9, 11. ☐*non* 1920 *Apatokephalus serratus* (Boeck) – Størmer, p. 10, Pl. 2:3. ☐*non* 1927 *Apatokephalus serratus* (Boeck) – Stubblefield & Bulman, p. 136, Pl. 4:7. ☐*non* 1931 *Apatokephalus* aff. *serratus* (Boeck) – Klouček, p. 367, Pl. 1:9, *non* Pl. 1:8. ☐*non* 1931 *Apatokephalus serratus* (Boeck) – Lake, pp. 122–123, Pl. 14:14. ☐*non* 1932 *Apatokephalus serratus* (Boeck) – Lake, p. 149, Pl. 19:1, 2. ☐1937 *Apatokephalus serratus* (Boeck) – Raymond, p. 1084. ☐*non* 1938 *Apatokephalus serratus* (Boeck) – Harrington, p. 169, Pl. 5:1–5. ☐1940 *Apatokephalus serratus* (Boeck) – Størmer, p. 139. ☐*non* 1954 *Apatokephalus serratus* (Boeck) – Wilson, p. 275, Pl. 27:1, 2, 13. ☐1956a *Apatokephalus serratus* (Boeck) – Tjernvik, pp. 204–206, Text-fig. 32A, Pl. 2:7, 8. ☐*non* 1957 *Apatokephalus serratus* (Boeck) – Harrington & Leanza, pp. 135–139, Fig. 56:1–10. ☐1959 *Apa-*

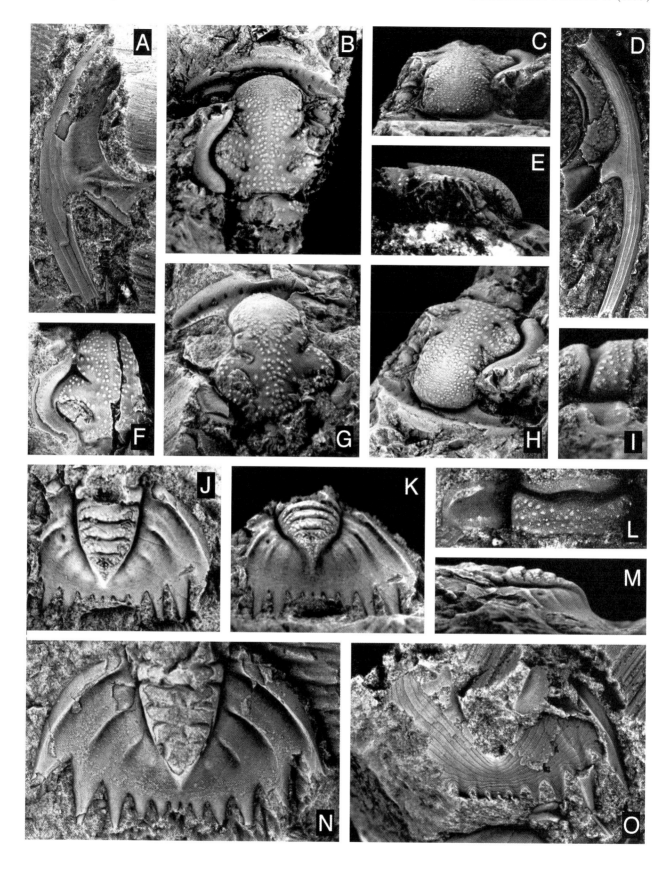

short, extending obliquely backwards from posterior extremity of eye socle. Narrow ridge extends from posterior extremity of eye socle to base of genal spine. Another short ridge extends from same base directly to eye socle. Anterior facial suture extends backwards, defining spindle-shaped anterior area of genal field, then turning round, defining narrow, pointed anterior part of border. Border furrow shallow, defining bevelled border with open-spaced terrace ridges.

Thoracic segment with wide (tr.) rachial ring, nearly half total width. Articulating half-ring narrow (sag.), tapering laterally (Fig. 42I). Median part of rachial furrow crescentic forward, merging posteriorly with rachial ring. Rachial ring covered with tubercles. Pleural field semi-rectangular with wide (sag.) pleural furrow defining two low, marginal ridges with small tubercles (Fig. 42L). Number of thoracic segments unknown.

Maximum width (tr.) of pygidium opposite posterior part of rachis, equals sagittal length (excluding marginal spines). Rachis long (sag.), tapering backwards, ending in short, narrow postrachial ridge, not reaching margin (Fig. 42N). Width of rachis nearly one-third of maximum width. It carries four rachial rings covered with tubercles. Median part projecting slightly backwards, forming tile-like structure (Fig. 42M). Pleural fields horizontal near rachis, sloping steeply to flat, narrow border. Four pleural furrows present, curving obliquely backwards from near rachis, parallel to outline of anterior margin, forming distinct terraces. Posterior pleural furrow extends to margin, outlining lateral marginal spine (Fig. 42N). Posterior pleural furrow faintly indicated. Lateral margin curving outwards, backwards from posterior part of rachis to point of maximum width. Posterior margin semielliptical

Fig. 42. Apatokephalus serratus (Boeck, 1838). □A. Dorsal view of left free cheek, showing facial suture; ×4, PMO 136.064/1. Found 16–19 cm above the base of the formation at Prestenga bus stop in Slemmestad, Røyken. Coll.: J.O.R. Ebbestad, 1992-09-28. □B, C, E, H. Dorsal, anterior, lateral and oblique anterolateral view of cranidium; ×5.5, PMO 121.635/2. Found 23–28 cm above the base of the formation at Vestfossen Railway st., Øvre Eiker. Coll.: J.O.R. Ebbestad, 1992-09-02. □D. Ventral view of left free cheek, showing doublure; ×5, PMO 121.648. Found 40–80 cm above base of the formation at Øvre Øren, Modum. Coll.: J.O.R. Ebbestad, 1992-08-26. □F. Dorsal view of central glabella, showing structures of the exoskeleton; ×5.5, PMO 1473. Vestfossen, Øvre Eiker. Coll.: W.C. Brøgger, 1879. □G. Dorsal view of nearly complete cranidium, showing anterior border (internal mould); ×4.5, PMO 83985/2. Bjørkåsholmen, Asker. Coll.: B.-D. Erdtmann, 1963. □I, L. Lateral and dorsal view of thoracic segment, with distinct granulation; ×10, PMO 121.635/1. 23–28 cm above base of the formation at Vestfossen railway st., Øvre Eiker. Coll.: J.O.R. Ebbestad, 1992-09-02. □J, K, M. Dorsal, posterior and lateral view of pygidium; ×6, PMO H2622. Vestfossen, Øvre Eiker. Coll.: W.C. Brøgger, 1879. Figured by Brøgger (1882, Pl. 3:7) and Fortey & Owens (1991, Fig. 12p). □N. Dorsal view of pygidium with distinct pleural spines; ×5, PMO 84194. Elnestangen, Asker. Coll.: F. Nikolaisen, 1967-06-01. Figured by Fortey & Owens (1991, Fig. 12q). □O. Dorsal view of pygidial doublure, showing terrace ridges; ×8.5, PMO S1153. Bygdøy Sjøbad, Oslo. Coll.: L. Størmer, 1918.

with twelve short, pointed spines, being proportionally smaller adrachially. Doublure wide with open-spaced terrace ridges (Fig. 42O).

Discussion. – Dean (1966, p. 343) noted that the reconstruction of this species published in the *Treatise on Invertebrate Paleontology* (Moore, 1959, p. O331, Fig. 243:2a, b) is highly inaccurate, both concerning proportions and concerning the structure of the cranidium and pygidium. Based on the material described herein, a new reconstruction is provided (Fig. 41A).

Fortey & Owens (1991, pp. 457–458) assigned material from Great Britain, formerly named *A. serratus*, to the new species *A. sarculum*. They drew attention to the structure of the cephalic sculpture, which differs markedly between the two. It can also be noted that *A. serratus* have more strongly geniculating S1 furrows positioned more marginally, shorter (tr.) S3 furrows, no apparent posterior fixed cheeks and no terrace ridges at the anterior border.

Most other species formerly referred to *A. serratus* are not conspecific. *A. serratus dubius* of Moberg & Segerberg (1906) is here considered to be *A. dubius* (Linnarsson, 1869); *A. serratus* of Lake (1931, 1932), Stubblefield & Bulman (1927) and Thomas *et al.* (1984) is now *A. sarculum* Fortey & Owens, 1991; *A. serratus* of Størmer (1920) is *A.* cf. *sarculum* Fortey & Owens, 1991; *A. serratus* of Klouček (1931) is now *A. asarkus* Vaněk, 1965; *A.* aff. *serratus* of Klouček (1931) is now *A.?* sp. *sensu* Mergl (1994); *A. serratus* of Harrington & Leanza (1957) is *A. tibicens* Přibyl & Vaněk, 1980; *A. serratus* of Wilson (1954) is now *A. wilsoni* Dean, 1966; *A. serratus schoriensis* of Poletaeva (*in* Petrunina *et al.* 1960) is *A. schoriensis* Poletaeva; *A. serratus serratus* of Poletaeva (*in* Petrunina *et al.* 1960) was referred to *A. serratus* by Burskij (*in* Bondarev *et al.* 1965). However *A. serratus sensu* Burskij (*in* Bondarev *et al.* 1965, 1966, 1970) differs from the Norwegian material in having denser granulation, 3S furrow obliquely backwards, and a narrow brim-like anterior border. It is here considered a different species. *A. serratus* of Liu (*in* Zhou *et al.* 1977) is now *A. latilimbatus* Peng, 1990.

Apatokephalus dubius (Linnarsson, 1869)

Figs. 41B, 43

Synonymy. – □1869 *Remopleurides dubius* n.sp. – Linnarsson, p. 69, Pl. 1:26. □1882 *Remopleurides dubius* (Linnarsson) – Brøgger, p. 127, Pl. 3, *non* Fig. 14. □1896 *Apatokephalus dubius* (Linnarsson – Brøgger, pp. 175, 180–184, *non* Fig. 5a, b. □1906 *Apatokephalus serratus* (Boeck) var. *dubius* – Moberg & Segerberg, pp. 88–89, Pl. 5:10. □*non* 1907 *?Apatokephalus serratus* (Boeck) var. *dubius* Moberg & Segerberg, Schmidt, p. 66, Pl. 3:13. □1914 *Apatokephalus dubius* (Linnarsson) – Walcott, p. 350. □*non*

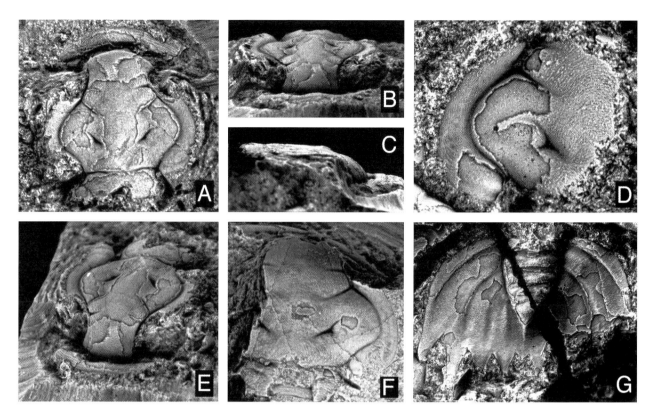

Fig. 43. Apatokephalus dubius (Linnarsson, 1869). □A–C, E. Dorsal, anterior, lateral and oblique anterolateral view of cranidium; ×5.5, PMO 1108/2. Engervik, Asker. Coll.: W.C. Brøgger, 1879. □D. Dorsal view of left palpebral lobe and glabella, showing structures of the exoskeleton; ×9, PMO 1472. Vestfossen, Øvre Eiker. Coll.: W.C. Brøgger, 1879. □F. Dorsal view of incomplete cranidium, showing glabellar furrows and structures of the exoskeleton; ×5.5, PMO 93629/4. Bjørkåsholmen, Asker. Coll.: J.F. Bockelie, 1968. □G. Dorsal view of pygidium, retaining most of the exoskeleton; ×5.5, PMO 977/1. Bjørkåsholmen, Asker. Coll.: V. Gaertner and T. Strand, 1928-05-13.

1938 *Apatokephalus dubius* Linnarsson – Harrington, p. 171, Text-fig. 5, Pl. 5:6–10. □1956a *Apatokephalus dubius* Linnarsson – Tjernvik, pp. 205–206. □*non* 1961 *Apatokephalus serratus* (Boeck) var. *dubius* – Balashova, p. 129, Pl. 2:3. □1991 *Apatokephalus dubius* (Linnarsson) – Fortey & Owens, p. 458.

Type material. – The cranidium figured by Linnarsson (1869, Pl. 1:26) from the Bjørkåsholmen Formation at Mossebo in Hunneberg, Västergötland, Sweden, has not been identified among available collections in the Museum of Evolution, Uppsala, or the Geological Survey of Sweden, Uppsala, and may be lost. The Swedish material is poor and insufficiently known, but until the type is recovered or a neotype is selected, the Norwegian material must suffice to define the taxon.

Norwegian material. – Four cranidia, two pygidia. Tables 13 and 14 show the measurements.

Remarks. – The diagnosis given by Linnarsson (1869, p. 69) is inadequate, and an emended diagnosis is provided here.

Emended diagnosis. – Posterior part of glabella wider (tr.) than tongue-shaped anterior glabellar lobe. Anterior border narrow (sag.), bevelled with terrace ridges. Test of glabella covered with fine, raised fingerprint-like pattern. Maximum width (tr.) of pygidium posterior of rachial termination. It has four pairs of pygidial spines and one additional pygidial spine placed medially.

Description. – *A. dubius* differs from *A. serratus* (Boeck, 1838) in having a proportionally wider (tr.) occipital ring,

Table 13. Cranidial measurements of *Apatokephalus dubius.*

Specimen	A	B	C	C1	C2	J	J1	J2	K	K1
1108/2	0.83	0.71	0.29	0.42	0.05	0.54	–	0.74	–	0.33
1472	–	–	–	0.46	–	–	–	0.98	–	–
93629/4	–	0.75	–	0.51	0.09	–	–	1.02	–	0.36
121.560	0.90	0.80	–	–	–	–	–	1.02	–	0.45

Table 14. Pygidial measurements of *Apatokephalus dubius.*

Specimen	X	X1	Y	Z	W
977/1	0.32	0.14	0.47	0.65	1.05

a less distinct occipital furrow, less distinct 1S furrows, bifurcated and positioned more adrachially, less distinct 2S and 3S furrows, positioned deeper adrachially, a proportionally narrower (tr.) frontal glabellar lobe, bulging at preglabellar field (Fig. 43B), less distinct palpebral lobe furrows, palpebral lobes tapering anteriorly, no preglabellar area, a narrow (tr.) border furrow defining a bevelled border carrying terrace ridges, and an exoskeletal pattern of fine, raised lines, resembling fine fingerprints (Fig. 43D).

The pygidium differs from the type species in having a less curved anterior margin defining a more quadratic outline, less well-defined pleural furrows, a more transverse posterior margin with only four pairs of pleural spines that are more triangular, an additional single spine positioned medially, and a slightly granulated exoskeleton (Fig. 43G).

Discussion. – The material described here is similar to *A. dubius* figured and described by Linnarsson (1869), and is considered conspecific. Tjernvik (1956a, p. 206) and Fortey & Owens (1991) also recognized *A. dubius* as a disitinct species and not a variant or subspecies of *A. serratus*.

Harrington (1938) described a species from Argentina as *A. dubius* (Linnarsson, 1869), but the cranidium differs markedly in having longer 1S furrows positioned closer to the margin, a shorter, more rounded frontal glabellar lobe, a wide preglabellar area, and a more narrow anterior border. The pygidium differs mainly in having a transverse anterior margin close to the rachis, and five pairs of larger pleural spines.

A. dubius is similar to species of *Eorobergia*, such as *E. marginalis* (Raymond, 1925) and *E. breviceps* (Raymond, 1925), in the distinction of the glabellar furrows, the structure of the test, and in lacking a preglabellar area. However, the pygidium is very much like that of *A. serratus*. Nikolaisen (1991, p. 51) believed that species of the *E. marginalis* group of trilobites developed from *Apatokephalus* or closely related species. *A. dubius* strengthens this assumption, but awaiting a revision of the genera of *Eorobergia* and *Apatokephalus*, the generic status of *A. dubius* remains unaltered.

Apatokephalus dactylotypos n.sp.

Figs. 41C, 44

Derivation of name. – From Greek *dactylos* meaning finger, and *typos* meaning mark, impression, referring to the fingerprint-like pygidial ornament.

Holotype. – Identified here; a cranidium (PMO 66831/2) from a dark limestone nodule near the base of the Bjørkåsholmen Formation at Skarra cross road in Vestfossen, Øvre Eiker, Norway.

Paratypes. – Selected here; a pygidium (PMO 144.351) found 9–12 cm above the base of the formation at Skarra Farm in Vestfossen, Øvre Eiker, and a free cheek (PMO 1240/4) found in Asker, Bærum. Both samples originate from dark limestone nodules near the base of the Bjørkåsholmen Formation.

Diagnosis. – A species of *Apatokephalus* which differs from the type species in having no ornament on the marginal areas of glabella between palpebral lobes. Front of glabella laterally expanded with slight median bulge. Anterior facial sutures diverging with 90° angles to glabella in front of anterior glabellar furrows. Pygidium with distinct densely spaced fingerprint-like ornament and fine tubercles.

Description. – Length (sag.) of cranidium nearly twice posterior width (tr.) of glabella. Median part of glabella slightly convex (Fig. 44D). Occipital furrow distinct, transverse, widening slightly (sag.) laterally. Preoccipital part of glabella expanded between palpebral lobes, almost as wide (tr.) as preoccipital part of glabella is long (sag.). Frontal glabellar lobe expanded laterally and bulging slightly medially (Fig. 44A). Marginal areas of expanded part without ornament. Rest of glabella covered with small, densely spaced tubercles anteriorly following general outline of frontal glabellar lobe. Preoccipital furrows situated opposite middle (sag.) of palpebral lobes and halfway (tr.) to median line. They are short, bifurcating, with the longer branch curving anteriorly. Middle glabellar furrows slightly curved, directed obliquely backwards, situated laterally opposite posterior part of palpebral lobe furrows. Frontal glabellar furrows transverse, situated marginally posteriorly of frontal glabellar lobe expansion. Palpebral lobes relatively wide (tr.) and long (sag.), slightly expanded posteriorly. Frontal facial sutures diverging with 90° angle to glabella opposite maximum width of frontal glabellar lobe. Preglabellar area wide (sag.).

Free cheek similar to that of type species, except genal field is proportionally larger, area between posterior margin and eye ridge is subquadratic, only one distinct ridge extends from base of genal spine, genal area with reticulated pattern, and terrace ridges on the border are more densely spaced (Fig. 44C).

Pygidium with four rachial rings (Fig. 44B). Width (tr.) of rachis subparallel to opposite third and fourth rachial ring before tapering backwards. Ending in terminal piece with distinct postrachial ridge and steep transition to postrachial area. Four pleural furrows present. Anterior pleural furrow with short, indistinct interpleural furrow near rachis. Posterior pleural furrow indistinct. Pleural furrows form steplike areas in otherwise gently curved pleural fields. Posterior margin and marginal spines unknown. Entire pygidium covered with densely spaced

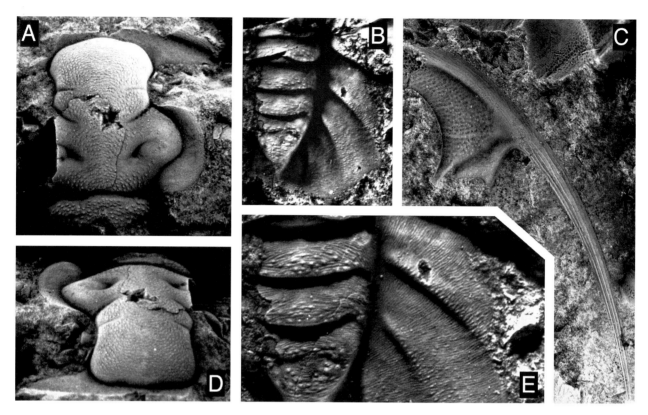

Fig. 44. Apatokephalus dactylotypos n.sp. □A, D. Dorsal and frontal view of holotype cranidium, showing palpebral lobe and granulation; ×8, PMO 66831/2. Dark limestone nodule 5–10 cm above base of the formation at Skarra, Vestfossen, Øvre Eiker. Coll.: G. Henningsmoen, 1954-07-10. □B, E. Dorsal view with detail of sculpture of incomplete paratype pygidium; B ×10; E ×19, PMO 144.351. Found in a dark limestone nodule 8–12 cm above the base of the formation at Skarra farm, Vestfossen Coll.: J.O.R. Ebbestad 1995-10-05. □C. Dorsal view of right free cheek; ×4, PMO 1240/4. Found in a dark limestone nodule 25–30 cm above base of the formation at Bjørkåsholmen, Asker. Coll. unknown, 1915.

fingerprint-like ornament (Fig. 44E). Posterior part of rachial rings with additional small tubercles.

Discussion. – This form is distinguished as a separate species, although it is known only from single specimens of a cranidium, a free cheek and a pygidium, respectively, from one stratigraphical position. The two other species of *Apatokephalus* in the Bjørkåsholmen Formation are represented by quite well-defined material from a different stratigraphical level. It could be argued that *A.* cf. *sarculum* Fortey & Owens, 1991, from the underlying strata of the Alum Shale Formation (see below), represent the same species, or that *A. dactylotypos* n.sp. represent a subspecies; they are certainly similar in many respects. Nevertheless, the differences are substantial enough to warrant a separation. The expansion and median bulging of the frontal glabellar lobe are not seen in *A.* cf. *sarculum*. The frontal facial sutures of the latter starts diverging in a gentle curve from just in front of the anterior glabellar furrows, and the lateral margins of the facial sutures are almost opposite the same glabellar furrows. In *A. dactylotypos* n.sp., the frontal facial sutures diverge with an almost 90° angle to glabella opposite the maximum width

of the frontal glabellar lobe. The frontal margin seems also to have a gentler curve, giving a slightly narrower (sag.) preglabellar area. In the expression of the glabellar furrows the differences are minor, except that the preoccipital furrows reach closer to the lateral margins, and that the anterior glabellar furrows in *A. dactylotypos* n.sp. are relatively further away from the anterior glabellar margin. The ornament in *A. dactylotypos* n.sp. is lacking at the marginal areas between the palpebral lobes and are more symmetrically arranged on the frontal glabellar lobe. However, this is a feature in which some variations are expected within species of *Apatokephalus*.

Table 15. Cranidial measurements of *Apatokephalus dactylotypos* n.sp.

Specimen	A	B	C	C1	C2	J	J1	J2	K	K1
66831/2	–	0.63	–	0.35	0.11	0.44	0.93	0.72	0.44	0.38

Table 16. Pygidial measurements of *Apatokephalus dactylotypos* n.sp.

Specimen	X	X1	Y	Z	W
144.351	0.28	0.23	0.57	0.68	–

The transverse anterior facial sutures, the lateral expansion in the front of the glabella, and the median bulge seen in *A. dactylotypos* n.sp. distinguish this species from most other species assigned to the genus. *A. tibicens* from Argentina (Přibyl & Vaněk 1980) and *A. sarculum* from Shropshire (Fortey & Owens 1991) are similar to *A. dactylotypos* n.sp. in the expression of the anterior facial sutures, but lack the lateral frontal glabellar expansion and a median bulge. The lateral glabellar furrows are also quite different, being short (tr.) and deeply inserted in *A. dactylotypos* n.sp.; sigmoidal and longer (tr.) in *A. tibicens* and *A. sarculum*. The pygidia of the three species differ in outline, definition of the pleural fields, lateral border spines, and ornamentation.

A slight lateral expansion of the frontal glabellar lobe resembling that of *A. dactylotypos* n.sp. can be seen in *A. levisensis* (Rasetti 1943). The position and direction of the anterior facial sutures are also very similar in the two species. However, the glabellar lobes and ornamentation distinguish the two. The pygidium of *A. levisensis* only has three rachial rings.

The lateral part of the frontal glabellar lobe in species like *A. poletaeva* Fedyanina, 1960, and *A. serratus serratus sensu* Poleteva (1960, see discussion of *A. serratus*), may be expanded, but the illustrations and preservational state of the fossils make it difficult to see this feature.

Tables 15 and 16 give measurements of two specimens.

Apatokephalus cf. *sarculum* Fortey & Owens, 1991

Figs. 41D, 45

Synonymy. – □1940 *Apatokephalus serratus* (Boeck, 1838) – Størmer, p. 10, Pl. 2:3. □1956a *Apatokephalus serratus* (Boeck, 1838) – Tjernvik, p. 205. □cf. 1991*Apatokephalus sarculum* n.sp., Fortey & Owens, p. 455, Fig. 12a–j, l, o.

Holotype. – A small, well-preserved, but incomplete dorsal exoskeleton (NMW 86.27G. 14), from the Arenaceous Member, Coundmoor Brook, Shropshire. Identified and figured by Fortey & Owens (1991, p. 455, Fig. 12a).

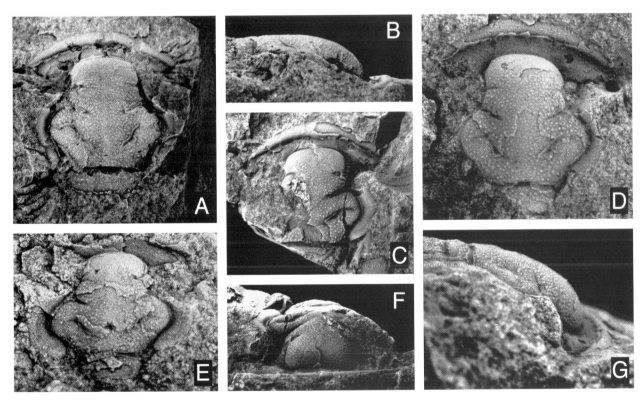

Fig. 45. Apatokephalus cf. *sarculum* Fortey & Owens, 1991. □A, B. Dorsal and lateral view of cranidium; ×3, PMO 20023. Found in limestone of the upper Alum Shale Formation at Vækerø, Oslo. Coll.: L. Størmer, 1920. □C, F. Dorsal and anterior view of cranidium; ×3, PMO 84162. Found in limestone at profile A (Størmer 1920), in the upper Alum Shale Formation at Vækerø, Oslo. Coll.: L. Størmer, 1921. □D, G. Dorsal and lateral view of cranidium; ×11, PMO 471. Found in limestone at profile A (Størmer 1920), in the upper Alum Shale Formation at Vækerø, Oslo. Coll.: L. Størmer, 1918–1919. Figured by Størmer (1920, Pl. 2:3a–b). □E. Dorsal view of small cranidium; ×11, PMO 97283. Found 20 cm above the Platypeltoides Limestone at profile A (Størmer 1920), in the upper Alum Shale Formation at Vækerø, Oslo. Coll.: J. Gjessing, 1974.

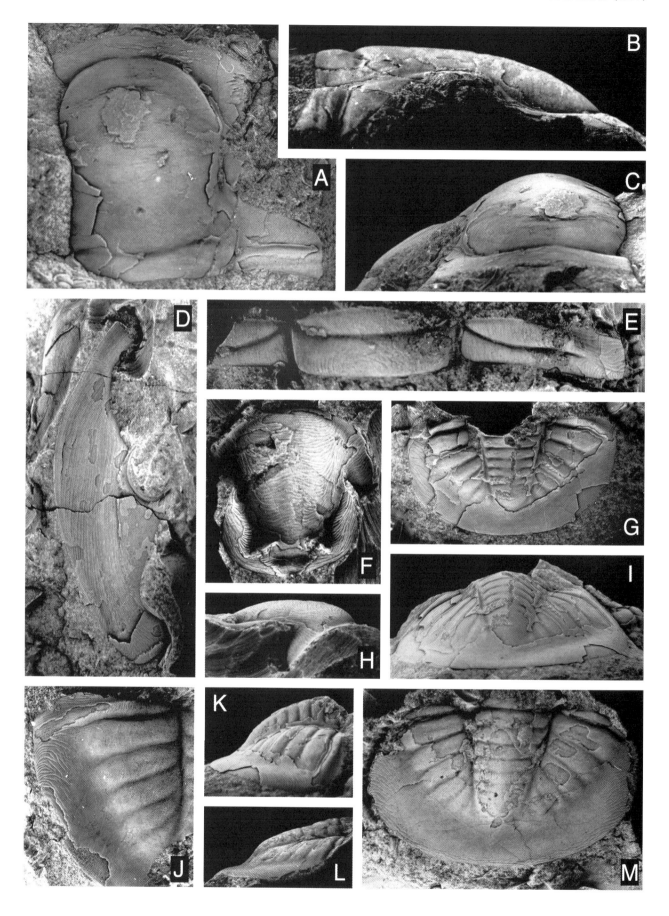

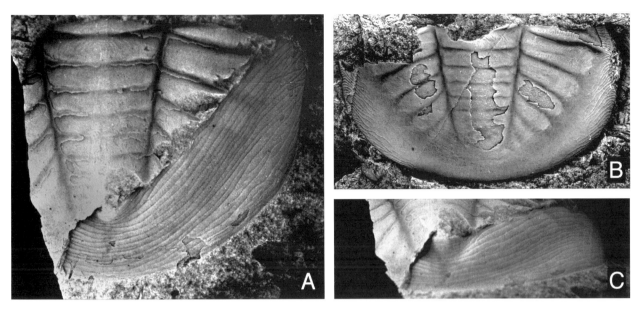

Fig. 48. Niobe (Niobe) insignis Linnarsson, 1869. □A, C. Dorsal and posterior view of pygidium, showing doublure; ×3, PMO 64123. Kutangen, Røyken. Coll.: L. Størmer, 1942-09-13. □B. Dorsal view of pygidium, showing pleural fields, posterior border and structures of the exoskeleton; ×2, PMO H2611. Vestfossen, Øvre Eiker. Coll.: W.C. Brøgger, 1879. Figured by Brøgger (1882, Pl. 2:1c).

Remarks. – The diagnosis given by Linnarsson (1869, p. 75) is inadequate. His illustration also deviates from subsequent descriptions and illustrations (Brøgger 1882; Moberg & Segerberg 1906; Holm 1901), mainly by showing a pygidium with a longer rachis and seven or eight pleural lobes. This was also noted by Tjernvik (1956a, p. 225). To this can be added that the marginal termination of the pleural lobes is indistinct on the original illustration (Linnarsson 1869, Pl. 2:36). Tjernvik regarded the common form found and described by the authors mentioned above as conspecific with *Niobe insignis* of Linnarsson (1869). That view is followed here. An emended diagnosis and description based on the Norwegian material must suffice until the holotype is found or a neotype selected.

Fig. 47. Niobe (Niobe) insignis Linnarsson, 1869. □A–C. Dorsal, lateral and anterior view of large cranidium; ×2, PMO 88639/2. Bjørkåsholmen, Asker. Coll.: G. Henningsmoen, 1958-11-16. □D. Dorsal view of left free cheek doublure; ×2, PMO 83906. Engervik, Asker. Coll.: G. Henningsmoen, 1959-03-15. □E. Dorsal view of thoracic segment, showing pleura and rachial ring; ×2, PMO 1341/2. Bjørkåsholmen, Asker. Coll. unknown, 1915. □F, H. Dorsal and lateral view of hypostome, showing anterior wing and structures of the exoskeleton; ×2, PMO 83732. Bjørkåsholmen, Asker. Coll.: G. Henningsmoen, 1959-01-03. □G, I, K. Dorsal, posterior and lateral view of pygidium, showing strong convexity; ×2.5, PMO 1334/1. Bjørkåsholmen, Asker. Coll. unknown, 1915. □J. Anterolateral margin of pygidium, showing terrace ridges; ×3.5, PMO 121.635/4. Found 23–28 cm above base of the formation at Vestfossen railway st., Øvre Eiker. Coll.: J.O.R. Ebbestad, 1992-09-02. □L, M. Lateral and dorsal view of pygidium; L ×2.5; M ×3, PMO 58720/1. Bjørkåsholmen, Asker. Coll. unknown, 1915.

Emended diagnosis. – Glabella with distinct bacculae, three pairs of glabellar furrows merely forming depressions. Anterior glabellar lobe slightly expanded, bluntly rounded, tapering anteriorly to merge with rounded front. Hypostome with only anterior lobe developed, and posterior margin slightly pointed. Pygidium with six rachial rings, and six pleural ribs with distinct marginal termination. Posterior margin slightly semielliptical, lateral margin nearly transverse.

Emended description. – Sagittal length of cranidium two-thirds of posterior width. Glabella wide (tr.) and long (sag.), occupying half the posterior width and nearly 90% of the sagittal length, weakly convex (tr.), slanting anteriorly, bluntly rounded in front. Posterior part expanded laterally into distinct bacculae, narrowing just posterior of palpebral lobes, then expanding slightly anterior of eyes. Occipital furrow wide (tr.), distinct, curving slightly backwards sagittally, transverse laterally, not reaching margin (Fig. 47A). Glabellar furrows indicated as depressions only. S1 furrows, short, transverse, positioned laterally midway between posterior end of palpebral lobes and occipital furrow. S2 furrows transverse, opposite anterior part of palpebral lobes. S3 furrows extending slightly obliquely anteriorly from anterior extremities of eyes. T-shaped depression present medially, midway between anterior extremity of eyes and preglabellar furrow. Anterior glabellar lobe tapering slightly forward to merge with rounded frontal part. Small median tubercle situated midway between posterior part of palpebral lobes and occipital furrow. Dorsal furrow indistinct. Posterior part of fixed cheeks subrectangular, making up one-third of sagittal length. Suture directed exsagittally at lateral margin. Posterior margin and posterior border furrow transverse. Palpebral lobes small, making up one-fifth of sagittal length, semicircular, positioned close to and opposite the middle of glabella. Anterior width of fixed cheek almost equal to sagittal length. Preglabellar field narrow (sag.), expanded anterolateral. Anterior facial suture diverging outwards, rounded anterolaterally to merge with smooth, semielliptical anterior margin.

Free cheek elongated (Fig. 47D). Genal fields smooth without distinct transition to border. Terrace ridges present marginally. Doublure wide (tr.) with distinct terrace ridges. Margin smooth, rounded, without genal spines. Genal angle semiangular.

Hypostome large, quadrangular. Median body slightly convex (sag., tr.), merging anterolaterally with anterior wings, tapering posteriorly to meet with pair of deep maculae (Fig. 47F, H). Pair of pits positioned medially at narrow (sag.) posterior border. Posterior margin slightly pointed sagittally. Lateral border wide (sag.) posterolaterally, converging anteriorly to merge with median body opposite the middle of the hypostome. Anterior margin semicircular. Transverse terrace ridges at median body, subparallel to margin laterally.

Thoracic segments long (tr.) and narrow (sag.), with transverse margins (Fig. 47E). Rachial ring makes up one-third of width (tr.), carrying circular terrace ridges medially. Pleural furrows curving slightly backwards diagonally from anterior margin reaching two-thirds of pleural length (tr.).

Anterior width (tr.) of pygidium slightly less than twice sagittal length. Rachis wide, its width (tr.) at anterior margin slightly less than one-third of anterior width, making up 80% of sagittal length, tapering slightly posteriorly. It shows six rachial rings, becoming progressively less distinct posteriorly. Terminal pice bluntly rounded, extending down to border with a concave transition (Fig. 47K, L). Pleural fields convex curving down to slightly less convex border. Six pleural ribs present, directed slightly obliquely backwards, becoming shorter, less distinct posteriorly, obtusely rounded laterally. Border wide (tr.) posterolaterally, narrow (sag.) posteriorly. Short terrace ridges present marginally, directed slightly obliquely backwards laterally, transverse posteriorly (Fig. 48B). Posterior margin slightly semielliptical, lateral margins almost sagittal anterior of rachial ring number six (Figs. 47G, 48A). Anterior margin transverse close to rachis, rounded laterally outside pleural ribs. Doublure wide (tr.), with open-spaced terrace ridges subparallel to margin (Fig. 48A).

Discussion. – *Niobe insignis* differs from stratigraphically younger species in having a hypostome with a slightly pointed posterior margin, and not the typical bilobate one. This was excellently demonstrated by Tjernvik (1956a, Text-fig. 36).

N. incerta from the early Arenig strata of Sweden is remarkably similar to *N. insignis*, differing mainly in having a wider (tr.) anterior glabellar lobe, S4 furrow depressions marginally just posterior of eyes, eight rachial rings and pleural ribs, and a hypostome with bilobate posterior margins. The similar morphology and the close stratigraphical and geographical positions indicate a very close relationship.

The pygidium figured by Balashova (1961) as *N. insignis* is, judging from the rather poor illustration, not conspecific. It differs markedly in having a more strongly tapering rachis without distinct rachial furrows, and too few pleural ribs.

Nielsen (1995) suggested that the variations expressed in Scandinavian material of *N. (N.) insignis*, among others, reflected unrecognized dimorphism. However, the material represents a fairly complete growth series, and an analysis of the measurements taken does not readily support dimorphism for this species. However, distinct variations in convexity of pygidia with similar dorsal morphology can be seen (Fig. 47I, K, L, M).

Subgenus *Niobe* (*Niobella*) Reed, 1931

Type species. – *Niobe homfrayi* Salter, 1866, pp. 143–144, Pl. 20:3–12, from lower Tremadoc slates at Penmorfa Church near Tremadoc, Wales; by original designation of Reed (1931).

Discussion. – Lisogor (1977), Shergold & Sdzuy (1984), Peng (1990) and Nielsen (1995) listed species belonging to *Niobella*. The distinction between *Niobe* and *Niobella* is discussed under the paragraph on the subgenus *Niobe*.

Niobe (*Niobella*) *obsoleta* Linnarsson, 1869

Fig. 49

Synonymy. – □1869 *Niobe obsoleta* n.sp. – Linnarsson, p. 75, Pl. 2:35. □1882 *Niobe obsoleta* Linnarsson – Brøgger, p. 66, Pl. 4:2. □1886 *Niobe obsoleta* Linnarsson – Brøgger, p. 49. □1906 *Niobe obsoleta* Linnarsson – Moberg & Segerberg, p. 95, Pl. 6:15, 16. □*non* 1906 *Niobe obsoleta* Linnarsson – von Post, Figs. 1, 2. □1956a *Niobella obsoleta* (Linnarsson) – Tjernvik, pp. 228–229, Text-fig. 37A, Pl. 5:1, 2. □1980 *Niobella obsoleta* (Linnarsson) – Tjernvik & Johansson, p. 202. □1984 *Niobella obsoleta* (Linnarsson) – Shergold & Sdzuy, p. 105.

Lectotype. – A pygidium (SGU K.V.A.H. VIII, 2) from the Bjørkåsholmen Formation at Mossebo in Hunneberg,

Table 18. Cranidial measurements of *Niobe* (*Niobella*) *obsoleta.*

Specimen	A	B	C	C1	C2	J	J1	J2	K	K1
1114/1	1.47	1.29	0.63	0.30	0.42	–	1.41	–	1.05	0.78
1214/4	1.60	1.40	–	–	–	2.00	1.20	–	1.00	0.80
1476/1	2.80	2.60	1.10	–	–	3.60	2.20	–	1.70	1.60
1478/4	1.90	1.70	0.70	0.40	–	–	1.40	1.60	1.40	1.10
83744/2	1.01	0.89	0.41	0.29	0.29	1.44	0.81	0.90	0.64	0.56
83870	1.80	1.60	0.65	0.50	0.50	2.15	1.30	1.50	1.10	1.00

Table 19. Pygidial measurements of *Niobe* (*Niobella*) *obsoleta.*

Specimen	X	X1	Y	Z	W	W1
H2612	0.33	0.18	0.63	0.78	1.47	1.20
H2620	0.54	0.28	0.89	1.19	1.92	1.61
S1118	0.27	0.14	0.38	0.48	0.90	0.75
S1119	0.30	0.18	0.62	0.75	1.08	0.90
S3012	0.90	0.50	1.70	2.30	3.40	3.00
1137/3	0.60	0.38	1.08	1.40	2.40	2.16
1269/1	0.18	0.09	0.32	0.42	0.72	0.60
1534/1	0.36	0.20	0.60	0.78	1.56	1.35
83742	0.18	0.09	0.28	0.36	0.66	0.48
83760/1	0.47	0.24	0.90	1.08	1.83	1.62
83762	0.19	0.10	0.30	0.39	0.84	0.72
83969	0.13	0.08	0.22	0.31	0.56	0.50
84119	0.08	0.04	0.17	0.22	0.37	0.35
84122	0.29	0.16	0.44	0.67	1.14	0.90

Västergötland, Sweden. Illustrated by Linnarsson (1869, Pl. 2:35). Selected and figured by Tjernvik (1956a, p. 229, Pl. 5:2).

Norwegian material. – One cephalon, nine cranidia, 23 pygidia, two hypostomes and three free cheeks. Tables 18 and 19 present measurements of some of the cranidia and pygidia, respectively.

Remarks. – The diagnosis given by Linnarsson (1869, p. 74) was emended by Tjernvik (1956a, p. 229), but without including the cranidium. The diagnosis is therefore slightly modified here.

Emended diagnosis. – Occipital furrow short (tr.). Anterior and posterior width (tr.) of the glabella about equal. Palpebral lobes large, anterior width (tr.) of cranidium not reaching beyond palpebral lobe width (tr.). Anterior margin slightly pointed sagittally. Pygidium semicircular with laterally effaced pleural furrows. Rachis with seven rachial rings; posterior rachial ring may be indistinct. Pleural fields with five or six pleural furrows with weak interpleural furrows. Paradoublure line close to margins of pleural fields, intersecting rachis at posterior rachial ring. Border outlined by indistinct border furrow which is subparallel to posterior and lateral margin.

Description. – Tjernvik (1956a) emphasized the pygidial characters in his description. Here the distinction between *Niobe* (*Niobella*) *obsoleta* and the coeval *Niobe* (*Niobe*) *insignis* are outlined.

Niobe (*Niobella*) *obsoleta* differs from *N.* (*Niobe*) *insignis* in having a shorter (tr.) occipital furrow, the median node positioned just posteriorly of eyes, area of glabella between eyes more constricted, a slightly shorter (sag.) anterior glabellar lobe, expanding more laterally with a more evenly rounded front of glabella, the bacculae not as well developed, the posterior margin directed slightly obliquely backwards, a more pointed junction between posterior margin and lateral part of posterior facial sutures, the palpebral lobes positioned slightly more anteriorly, a stronger diverging anterior part of facial sutures, and more pointed (sag.) anterior margin. The free cheeks differ in having a more distinct border. The only two hypostomes known of *Niobe* (*Niobella*) *obsoleta* (PMO I 32, PMO1272) both come from dark nodules at the base of the Formation. The hypostome has proportionally narrower (tr.) lateral borders and wider (sag.) posterior borders, with a slightly trapezoidal outline.

The pygidium of *Niobe* (*Niobella*) *obsoleta* differs from that of *N.* (*Niobe*) *insignis* in having a width/length ratio of slightly more than half, having narrower rachis with seven rachial rings, five or six less distinct pleural ribs with faint indications of interpleural furrows and without distinct termination, paradoublure lines that are close to the margins of the pleural fields, a more horizontal border with nearly equal width posterior and laterally, a more semicircular outline of the margin. Generally *N.* (*Niobella*) *obsoleta* is the smaller species of the two.

Discussion. – *Niobe* (*Niobella*) *obsoleta* occurs in the dark limestone nodules near the base of the formation, and is probably confined to this level in the Norwegian strata. Specimens that occur in the succeeding limestone beds of the Bjørkåsholmen Formation were earlier included in *Niobe* (*Niobella*) *obsolete* but are here described as *Niobe* (*Niobella*) *eudelopleura* n.sp (see below).

The frontal glabellar lobe of *Niobe* (*Niobella*) *obsoleta* is shorter (sag.) and more rounded than that of *Niobe* (*Niobella*) *eudelopleura* n.sp. The anterior border is wider laterally in the latter species, and the anterior width (tr.) probably extends slightly beyond the width (tr.) of the palpebral lobes. Posterior fixed cheeks are proportionally larger (sag.), owing to a less steep inclination of the preocular suture.

Pygidia of *Niobe* (*Niobella*) *obsoleta* encountered in the dark limestone nodules (Fig. 49F–K) are morphologically very close to the lectotype. The pygidia encountered in the grey limestone beds are generally larger than the largest specimen found in the nodules. Additionally, the interpleural furrows are more pronounced, and the paradoublure line is further away from the margins of the pleural fields and intersects with the sixth rachial ring instead of the posterior, as in *Niobe* (*Niobella*) *obsoleta*. The lateral margins are also wider (tr.) in the new species, while the posterior border is narrow.

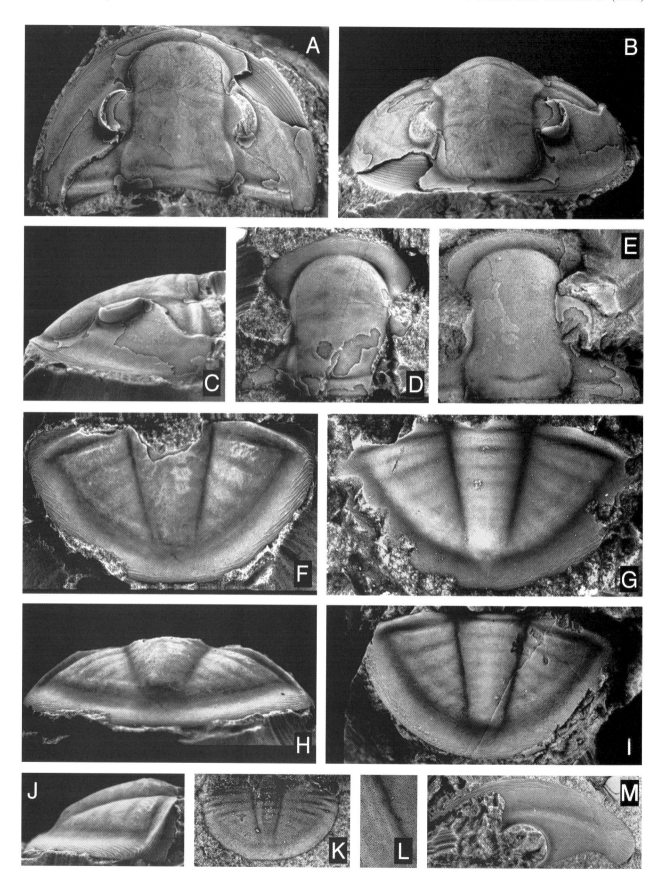

The bacculae of *Niobe* (*Niobella*) *obsoleta* are fairly prominent, but it can be seen that the bacculae increase in size from the Cambrian *Niobe* (*Niobella*) *primaeva* Westergård, 1909, to the Arenig *Niobe* (*Niobella*) *bohlini*. In the succeeding *Niobe* (*Niobella*) *imparilimbata* (Bohlin, 1955), the bacculae are even more distinct. The frontal glabellar lobe varies greatly in species of *Niobe* (*Niobella*). A slightly laterally expanded, well-rounded lobe is seen in *Niobe* (*Niobella*) *obsoleta*, while species like *Niobe* (*Niobella*) *imparilimbata* Bohlin, 1955, and *Niobe* (*Niobella*) cf. *plana* Nielsen, 1995, have a more truncated, trapezoidal outline of the frontal glabellar lobe, resembling that of *Gog* Fortey, 1975 (e.g., *Gog explanatus* (Angelin, 1854)).

The pygidia of Scandinavian Upper Cambrian and Lower Ordovician species of *Niobe* (*Niobella*) are very similar (see Tjernvik 1956a and Henningsmoen 1958). The Cambrian *Niobe* (*Niobella*) *primaeva* Westergård, 1909, has somewhat more distinct pleural furrows, while the Arenig *Niobe* (*Niobella*) *bohlini* (Tjernvik, 1956), has less distinct pleural furrows. Tjernvik & Johansson (1980) found that the pygidial doublure of *Niobe* (*Niobella*) *obsoleta* is narrow.

Fig. 49. Niobe (*Niobella*) *obsoleta* Linnarsson, 1869. □A–C. Dorsal, frontal and lateral view of complete cephalon; ×3, PMO 83870. Found in a dark limestone nodule 25–30 cm above base of the formation at Bjørkåsholmen, Asker. Coll.: G. Henningsmoen, 1958-09-11. □D. Dorsal view of cranidium, showing structures of the exoskeleton; ×2.5, PMO 1114/1. Found in a dark limestone nodule at Engervik, Asker. Coll.: W.C. Brøgger, 1879. □E. Dorsal view of cranidium, showing palpebral lobe; ×4.5, PMO 83744/2. Found in a dark limestone nodule 25–30 cm above base of the formation at Bjørkåsholmen, Asker. Coll.: G. Henningsmoen, 1961-10-13. □F, H, J. Dorsal, posterior and lateral view of small, nearly complete pygidium with distinct marginal terrace ridges; ×12, PMO 83742. Found in a dark limestone nodule 25–30 cm above base of the formation at Bjørkåsholmen, Asker. Coll.: G. Henningsmoen, 1961-10-13. □G. Dorsal view of small pygidium showing the border and distinction of the pleura and rachial furrows; ×11, PMO 1269/1. Found in a dark limestone nodule 25–30 cm above base of the formation at Bjørkåsholmen . Coll. unknown, spring 1915. □I. Dorsal view of large pygidium showing the border and distinction of the pleura and rachial furrows; ×5, PMO H2612. Found in a dark limestone nodule at Vestfossen. Coll.: W.C. Brøgger, 1879. □K. Latex replica from external mould of meraspid stage pygidium; ×18, PMO 84119. Found in a dark limestone nodule 14–17 cm above base of the formation at Sjøstrand, Asker. Coll.: G. Henningsmoen, 1959-03-01. □L. Detail of dorsal exoskeleton of free cheek, showing densely spaced pits; ×5, PMO 1202/2. Found in a dark limestone nodule at Bjørkåsholmen, Asker. Coll. unknown, 1915. □M. Dorsal view of right free cheek, showing marginal terrace ridges and facial suture; ×4, PMO 1266/2. Found in a dark limestone nodule at Bjørkåsholmen, Asker. Coll. unknown, 1915.

Niobe (*Niobella*) *eudelopleura* n.sp.

Figs. 50A–L

Synonymy. – □1906 *Niobe obsoleta* Linnarsson – Von Post, Figs. 1, 2. □1956a *Niobella* sp. aff. *obsoleta* (Linnarsson) no. 1 – Tjernvik, pp. 235.

Derivation of name. – From Greek *eudelos*, very clear, and *pleura*, side, referring to the pleural and interpleural furrows on the pygidium, which are clearly expressed on internal moulds, less so when the exoskeleton is preserved.

Holotype. – A pygidium (PMU Vg246), found at 46–52 cm in profile 1 at Stenbrottet in Västergötland, Sweden. Discussed by Tjernvik (1956, p. 235). Identified and illustrated here (Fig. 50B–D)

Paratypes. – A cranidium (Fig. 50A, PMO 136.072) found at Færdenveien in Klekken, Ringerike, Norway, in the interval 73–84 cm above the base of the formation. A pygidium (Fig. 50J, PMO 36012) from Ødegården in Snertingdalen, Norway. A pygidium (Fig. 50K, PMO 63160/1) from Bjørkåsholmen Asker, Norway.

Norwegian material. – Ten cranidia, 62 pygidia, and one hypostome. Tables 20 and 21 present measurements of some of the cranidia and pygidia, respectively.

Diagnosis. – Glabella elongated (sag.), with trapezoidal frontal glabellar lobe. Anterior border wide (tr.), with strongly divergent anterior facial sutures. Pygidium with seven rachial rings. Rachial furrows slightly undulating, not reaching across rachis. Pleural fields with four or five pleural furrows becoming indistinct posteriorly. Interpleural furrows distinct. Paradoublure line subparallel to lateral margin, about half-way out on the pleural fields, intersecting rachis at sixth rachial ring. Lateral pygidial border broad; posterior pygidial border narrow.

Description. – Sagittal length of cranidium four-fifths of posterior width (tr.). Glabella elongated; width approximately half of sagittal length, profile low, slanting gently anteriorly (Fig. 50A, E, F). Frontal glabellar lobe expanded in front of palpebral lobes, outline trapezoidal. Posterior part of glabella expanded laterally into distinct bacculae not reaching posterior margin. Occipital furrow short (tr.), distinct, slightly crescentic posteriorly. Glabellar furrows indicated as faint depressions halfway (exsag.) to median part of glabella. S1 furrows positioned opposite anterior part of bacculae. S2 furrows positioned opposite middle (sag.) of palpebral lobes. S3 furrows positioned opposite anterior part of palpebral lobes. Two median depressions present close to front of glabella. Median tubercle opposite S1 furrows. Posterior part of fixed

Discussion. – Fortey & Chatterton (1988) considered the Family Ceratopygidae to be a sister group of the family Asaphidae.

The genus includes eight species: *Ceratopyge forficula* (Sars, 1835) from the lower and upper Tremadoc of Norway and Sweden; *Ceratopyge acicularis* (Sars & Boeck, 1838) from the upper Tremadoc of Norway and Sweden; *C. latilimbata* Moberg & Segerberg, 1906, from the upper Tremadoc of Sweden; *C. forficuloides* Harrington & Leanza, 1957, from the lower and upper Tremadoc of Argentina; *C. transversa* Lu *in* Zhou *et al.*, 1977, from the Tremadoc of China; *C. elongata* Chang & Fan, 1960, from the Tremadoc of China; and *Ceratopyge* sp. from the Alum Shale and Bjørkåsholmen formations in Norway. *C. forficula sensu* Balashova (1961) from the Tremadoc of Kazakhstan, is here regarded as a different species.

The large, cosmopolitan genus *Proceratopyge* Wallerius, 1895, from upper Middle Cambrian and Upper Cambrian, shows the closest affinity to *Ceratopyge*. The pygidial spines of this genus are the continuation of the first segment only, whereas the pygidial spines in *Ceratopyge* are formed from the coalescence of distal parts of first and second segments.

The concept of the type species has for long comprised the two subspecies or variants *C. forficula forficula* (Sars, 1835) and *C. forficula acicularis* (Sars & Boeck, 1838). This confusion is unfortunate and is emended here. The stratigraphically older *C. forficula* (Sars, 1835) represents the true type species and is considered a distinct species separated from *C. acicularis* (Sars & Boeck, 1838) found in the Bjørkåsholmen Formation only. In addition, a third species recognized in the Norwegian strata, *Ceratopyge* sp. is contemporaneous with the two former species, but is only present at the very base of the range of *C. acicularis*. Fig. 55 presents reconstructions of the three species of *Ceratopyge* in the Lower Ordovician strata of Norway.

Ceratopyge forficula (Sars, 1835) was originally described from the upper part of the Alum Shale Formation, formerly Ceratopyge Shale (3aα–β), including forms having long or short pygidial spines of varying thickness (Sars 1835, Fig. 1d–f). The forms with short spines were later distinguished as *C. acicularis* by Sars & Boeck (*in* Boeck 1838, p. 141). The differentiation was based on the pygidial spines, stated to be shorter and narrower in *C. acicularis* and shaped like a harp in *C. lyra*. Brøgger (1882) chose to distinguished the three species as subspecies of *C. forficula*, with *C. forficula acicularis* as the form appearing in the Bjørkåsholmen Formation (Ceratopyge Limestone). Subsequent authors followed this usage, and a concept of *Ceratopyge forficula* based on morphological characters from the variants emerged. However, *C. lyra* was discarded as a variant by Størmer (1940, p. 140).

The variant *C. forficula lyra* from the Alum Shale Formation is here considered conspecific with *C. forficula*. Størmer (1940, p. 140) claimed that the type specimen

described and illustrated by Brøgger (1882, p. 124, Pl. 3:21) was missing. However, specimen PMO 56199 figured by Størmer (1940, Pl. 1:11) is probably the missing specimen, as indicated by the missing end of the right pygidial spine, the thin median line on the right spine, and the slight, but distinct, outward curve of the distal tip of that spine (most certainly a result of preservation). The specimen is reillustrated here (Fig. 57J).

The three species of *Ceratopyge* apparently have two morphological variants each (see Fig. 56). There are some indications that each species includes forms with long and shorth pygidial spines. The length of the pygidial spines was earlier claimed to be a distinguishing feature between the species (Gjessing 1976b), but it is evident that the length varies within all forms. This is likely to be related to dimorphism.

Ceratopyge forficula (Sars, 1835)

Figs. 55A–57

Synonymy. – ☐1835 *Olenus forficula* n.sp. – Sars, p. 332, Pl. 8:1a–f. ☐1838 *Trilobites forficula* (Sars) – Boeck, p. 141. ☐1838 *Trilobites lyra* n.sp. – Sars & Boeck mscr., Boeck, p. 141. ☐1847 *Ceratopyge forficula* (Sars) – Hawle & Corda, p. 161, Pl. 7:81. ☐1865 *Ceratopyge forficula* (Sars) – Kjerulf, p. 2, Fig. 9a, b. ☐1869 *Ceratopyge forficula* (Sars) – Linnarsson, p. 71. ☐1882 *Ceratopyge forficula* (Sars) – Brøgger, p. 123, Pl. 3:16–20, 22, *non* Figs. 15, 22. ☐1882 *Ceratopyge forficula* var. *lyra* (Boeck) – Brøgger, p. 124, Fig. 21. ☐1896 *Ceratopyge forficula* (Sars) – Koken, p. 19, Fig. 12:1, 2. ☐*non* 1901 *Ceratopyge forficula* (Sars) – Holm, p. 33, Fig. 28. ☐*non* 1906 *Ceratopyge forficula* (Sars) – Moberg & Segerberg, p. 85, Pl. 5:2–5. ☐1922 *Ceratopyge forficula* (Sars) – Raymond, pp. 207–208. ☐*non* 1940 *Ceratopyge forficula* (Sars) – Regnéll, pp. 88–89, Pl. 1:1–3. ☐1940 *Ceratopyge forficula* (Sars) – Størmer, pp. 126–127, Text-fig. 3:1d–f, Pl. 1:8–11. ☐1940 *Ceratopyge forficula* var. *lyra* (Boeck) – Størmer, p. 140. ☐1952 *Ceratopyge forficula* (Sars) – Skjeseth, Pl. 4:10, 11, 15, *non* Fig. 14. ☐1957 *Ceratopyge forficula* (Sars) – Harrington & Leanza, p. 185. ☐1959 *Ceratopyge forficula* (Sars) – Harrington *et al. in* Moore, p. O363, Fig. 271:1a, *non* Fig. 271:1b. ☐*non* 1961 *Ceratopyge forficula* (Sars) – Balashova, p. 130, Pl. 3:5–7. ☐*non* 1974 *Ceratopyge forficula* (Sars) – Balashova, pp. 489–489, Figs. 1–5. ☐1976b *Ceratopyge forficula forficula* (Sars) – Gjessing, pp. 134–141, Fig. 1B, C. ☐*non* 1989 *Ceratopyge forficula* (Boeck) – Bruton *et al.*, p. 237, Fig. 4:3, 6.

Lectotype. – A cranidium (PMO H2690) from the upper part of the Alum Shale Formation (Tremadoc), formerly Ceratopyge Shale (3aβ), in Oslo, Norway. Figured by Sars (1835, Pl. 8:1a). Selected and refigured by Størmer (1940, p. 127, Pl. 1:9).

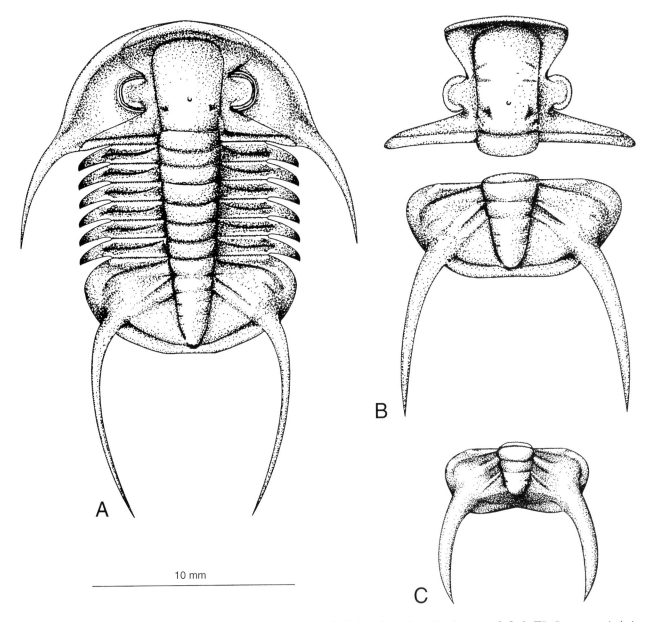

Fig. 55. Comparison of *Ceratopyge* species of the upper Alum Shale and Bjørkåsholmen formations. □A. *Ceratopyge forficula.* □B. *Ceratopyge acicularis.* □C. *Ceratopyge* sp. Scale bar 1 cm.

Remarks. – Størmer (1940, p. 127) incorrectly referred the museum catalogue number PMO 42690 to the lectotype specimen. His figure, however, is identical to the specimen with the number PMO H2690. The museum catalogue number PMO 42690 refers to a Silurian pentameride.

Material and occurrence. – Five cranidia and ten pygidia found in shale, and 20 cranidia, nine free cheeks, two hypostomes and 22 pygidia found in limestone in the shale unit. This material studied here comes exclusively from the upper part of the Alum Shale Formation, and this is probably also likely to represent the range of the species. Tables 22 and 23 present measurements of the cranidia and pygidia, respectively.

Diagnosis. – Preglabellar field narrow (sag.) and short (tr.), anterior facial suture straight. Pygidium with four rachial rings and transverse posterior margin. Adrachial flanks of the pygidial spines extend from posterior margin opposite rachial termination.

Description. – Sagittal length of cranidium nearly half of posterior width. Glabella with almost parallel sides making up 93% of sagittal length (Fig. 57B, C). Glabella slightly convex (tr.) medially, sloping steeply laterally,

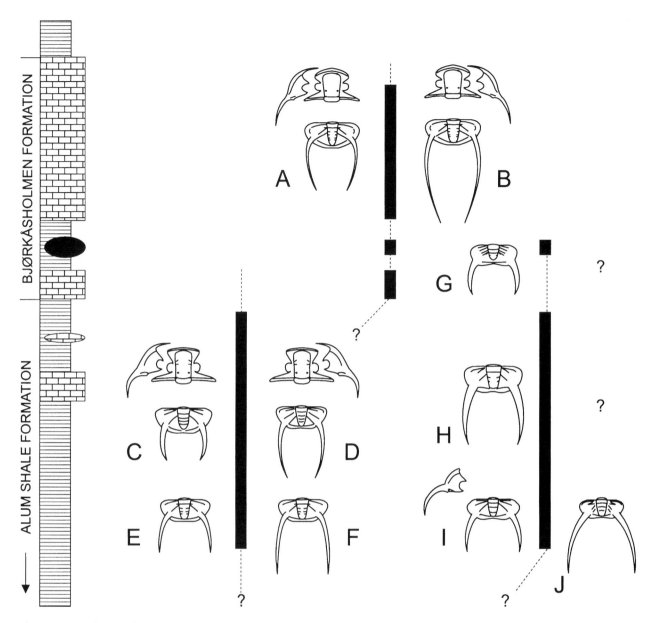

Fig. 56. Diagram showing the vertical distribution of *Ceratopyge*, indicating possible relationships between different morphotypes. *C. acicularis*: □A. Cranidium based on PMO S1334, free cheek based on PMO 20061/2, pygidium based on PMO S1348/1. □B. Cranidium and free cheek the same as above, pygidium based on PMO 121.578/1. *C. forficula*: □C. Cranidium based on PMO S996, free cheek based on PMO 704, pygidium based on PMO 543. □D. Cranidium and free cheek the same as above, pygidium based on PMO 93658. □E. Pygidium based on PMO 23644. □F. Pygidium based on PMO H2690b. *C. sp.*: □G. Pygidium based on PMO 1272/1. □H. Pygidium based on PMO 33142. □I. Free cheek based on PMO 84205, pygidium based on PMO 56193c. □J. Pygidium based on PMO 56193b and PMO 56194h.

sloping steeply to narrow (sag.) anterior glabellar field with almost transverse, bluntly truncated front (Fig. 57A–C). Occipital ring wide (sag.), making up one-fourth of posterior width, curving slightly crescentic backwards, defined anteriorly by transverse occipital furrow. Faint lateral depressions present at the basal baccula posterior of S1 furrows. S1 furrows deep, positioned laterally opposite posterior part of palpebral lobes. S2 glabellar furrows distinct, transverse, situated opposite ante-

rior extremities of palpebral lobes. S3 furrows as faint transverse indications slightly anterior of S2 furrows, directed obliquely forward (Fig. 57A, B). Posterior part of fixed cheek wide (tr.), spindle-shaped, curving only slightly laterally, with nearly transverse posterior margin. Posterior border furrow transverse, shallow, but distinct. Palpebral lobes tongue-shaped, short (sag.). Width (tr.) at eyes equals sagittal length. Anterior extremities of palpebral lobes situated just anterior to middle of gla-

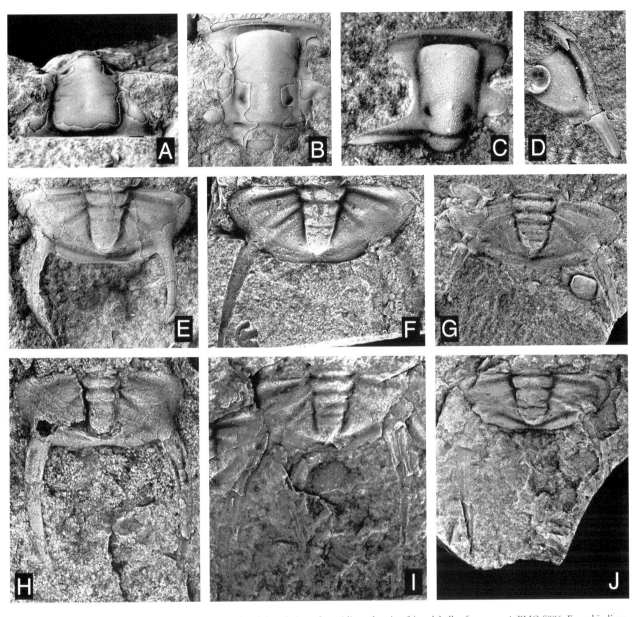

Fig. 57. Ceratopyge forficula (Sars, 1835). □A, B. Frontal and dorsal view of cranidium showing faint glabellar furrows; ×4, PMO S996. Found in limestone of the upper part of the Alum Shale Formation at Vækerø, Oslo. Coll.: L. Størmer, 1919. Figured by Gjessing (1976, Pl. 2:C). □C. Dorsal view of small cranidium with the exoskeleton preserved; ×8, PMO709. Found in limestone of upper part of the Alum Shale Formation at Fødselsstiftelsen in Stensberggaten, Oslo. Coll.: J. Kiær 1911. □D. Dorsal view of right free cheek; ×3 , PMO 704. Found in limestone of the upper part of the Alum Shale Formation at Fødselstiftelsen in Stensberggaten, Oslo. Coll.: J. Kiær, 1911. □E. Dorsal view of pygidium; ×3.5, PMO 543. Found in limestone of the upper part of the Alum Shale Formation at St. Olavsgate, Oslo. Coll.: T. Münster. □F. Dorsal view of pygidium ×3.5, PMO 23644. Found in shale of the upper part of the Alum Shale Formation at Ruseløkken, Oslo. Coll. unknown. □G. Dorsal view of pygidium with short pygidial spines; ×4, PMO 56194c. Found in shale of the upper part of the Alum Shale Formation at Ruseløkkveien, Oslo. Coll.: M. Sars. □H. Dorsal view of pygidium. Found in limestone of the upper part of the Alum Shale Formation at the road by Nedre Hvattum farm in Gran, Hadeland. Coll.: T. Münster, 1893-08-30. □I. Dorsal view of pygidium; ×4, PMO 42690b. Found in shale of the upper part of the Alum Shale Formation at Ruseløkken, Oslo. Coll.: M. Sars. Figured by Størmer (1940, Pl. 1:10). □J. Dorsal view of pygidium; ×4.5, PMO 56199. Shale of the upper Alum Shale Formation, Ruseløkken, Oslo. Coll. unknown. Figured by Brøgger (1882, Pl. 3:21) and Størmer (1940, Pl. 1:11).

bella. Anterior facial suture diverging at an angle of approximately 30° to exsagittal line, turning sharply inwards just opposite anterolateral corners of glabella, to merge with slightly curved anterior margin. Border furrow indistinct, defining almost horizontal, bevelled border. Border attenuated laterally.

Free cheeks elongate (Fig. 57D) Genal fields wide (tr.) and convex (tr.), curving down to narrow (tr.) border. At genal angle, genal field continues into long, broad genal spine, distinctly curved, not as continuation of semicircular cephalic margin. Suture opisthoparian.

Hypostome and thorax unknown.

Table 22. Cranidial measurements of *Ceratopyge forficula*.

Specimen	A	B	C	C1	C2	J	J1	J2	K	K1
S994	1.05	0.98	0.38	–	–	–	1.05	–	0.57	0.60
S995/1	0.61	0.54	0.24	0.19	0.16	–	0.68	–	0.28	0.33
S996	1.09	0.98	0.42	–	–	–	1.11	–	0.57	0.60
S1002	0.34	0.30	–	–	–	0.54	0.36	–	0.17	0.17
706	0.78	0.68	0.30	0.23	0.17	1.20	–	0.54	0.30	0.33
709	0.39	0.34	0.17	0.11	0.09	0.66	0.42	0.41	0.19	0.22

Table 23. Pygidial measurements of *Ceratopyge forficula*.

PMO	X	X1	Y	Z	W
S995/2	0.15	0.07	0.21	0.24	0.58
S995/3	0.06	0.05	0.14	0.17	0.42
S997/1	0.18	0.12	0.30	0.34	0.80
S998	0.18	0.10	0.27	0.34	0.74
S999	0.10	0.10	0.16	0.19	0.48
S1000	0.21	0.13	0.33	0.36	0.90
543	0.30	0.19	0.49	0.63	1.35
544	0.30	0.17	0.55	0.57	1.23
23644	0.24	0.13	0.40	0.43	1.01
33142	0.30	0.18	0.53	0.62	1.35
42690b	0.29	0.19	0.43	0.52	1.25
56193a	0.15	0.08	0.25	0.29	0.68
56194c	0.28	0.10	0.37	0.48	1.30
56194g	0.21	0.14	0.30	0.36	0.92
56199	0.30	0.18	0.38	0.42	1.14
93657	0.24	0.17	0.38	0.45	0.98
93658	0.22	0.12	0.33	0.41	0.90

Sagittal length of pygidium slightly less than half posterior width. Rachis narrow (tr.) and long (sag.), tapering slightly backwards to bluntly rounded rachial termination, curving steeply down to lateral border. Rachis with four rachial rings, anterior ring well defined, posterior rings indicated by faint rachial furrows or lateral depressions only (Fig. 57E, F). Dorsal furrow distinct. Pleural field horizontal adrachially, curving steeply down to border laterally and posteriorly. Pair of wide (tr.) and deep pleural furrows extends obliquely backwards from anterior margins of rachis, attenuating laterally. Posterior border narrow medially, wider (sag.) close to pygidial spines. Posterior margin nearly transverse, with slight median curve directed anteriorly. Anterior margin transverse, defining raised anterior border. Long, slender pygidial spines extending backwards as coalescence of pleural field embraced by first and second rachial ring, gently curved at base straight at posterior extremities. Spines extending slightly wider than anterior width, being 2–3 times longer than sagittal length of pygidium. Their adrachial flank protrudes from margin opposite rachial termination.

Discussion. – Balashova (1961, 1974) assigned a Tremadoc *Ceratopyge* from Kazakhstan to the *C. forficula*. They appear very similar, except that the in the Kazakhstan specimens the tubercle of the glabella is positioned closer to the occipital ring, the lateral parts of the posterior facial sutures are less pointed, the anterior facial sutures are less divergent, the pleural furrows of the thoracic segments are more distinct, the pointed ends of the pleurae are longer (tr.), the pygidial spines are straighter, reaching beyond the posterior width of the pygidium (might relate to state of preservation), and the pleural furrows are deeper incised. The Kazakhstanian specimens are here regarded as not conspecific with *C. forficula*.

Ceratopyge acicularis (Sars & Boeck, 1838)

Figs. 55B, 56, 58, 59

Synonymy. – ☐1838 *Trilobites acicularis* n.sp., Sars & Boeck mscr. – Boeck, p. 141. ☐1847 *Ceratopyge forficula* (Sars) – Hawle & Corda, p. 161, Pl. 7, *non* Fig. 81. ☐1865 *Ceratopyge forficula* (Sars) – Kjerulf, p. 2, *non* Fig. 9a, b. ☐1869 *Ceratopyge forficula* (Sars) – Linnarsson, p. 71. ☐1882 *Ceratopyge forficula* var. *acicularis* (Boeck) – Brøgger, p. 124, Pl. 3:15, 22. ☐1896 *Ceratopyge forficula* (Boeck) – Koken, p. 19, *non* Fig. 12:1, 2. ☐1901 *Ceratopyge forficula* (Sars) – Holm, p. 33, Fig. 28. ☐1906 *Ceratopyge forficula* (Sars) – Moberg & Segerberg, p. 85, Pl. 5:2–5. ☐1922 *Ceratopyge forficula* (Sars) – Raymond, pp. 207–208. ☐1940 *Ceratopyge forficula* (Sars) – Regnéll, pp. 88–89, Pl. 1:1–3. ☐*non* 1940 *Ceratopyge forficula* (Sars) – Størmer, p. 127, Pl. 1:8–11. ☐1940 *Ceratopyge forficula* var. *acicularis* (Sars & Boeck) – Størmer, p. 140. ☐1952 *Ceratopyge forficula* (Sars) – Skjeseth, Pl. 4:14, *non* Figs. 10, 11, 15. ☐*non* 1957 *Ceratopyge forficula* (Sars) – Harrington & Leanza, p. 185. ☐1957 *Ceratopyge forficula* var. *acicularis* (Sars & Boeck) – Harrington & Leanza, p. 185. ☐1959 *Ceratopyge forficula* (Sars) – Harrington *et al. in* Moore, p. 0363, Fig. 271:1b *non* Fig. 271:1a. ☐*non* 1961 *Ceratopyge forficula* (Sars) – Balashova, p. 130, Pl. 3:5–7. ☐*non* 1974 *Ceratopyge forficula* (Sars) – Balashova, pp.

Fig. 58. Ceratopyge acicularis (Sars & Boeck, 1838). ☐A. Dorsal view of cranidium retaining most of the exoskeleton; ×4, PMO 93499. Bjørkåsholmen, Asker. Coll.: J.F. Bockelie, 1966-09-09. ☐B, E, F. Dorsal, anterior and lateral view of cranidium, showing palpebral lobes and glabellar structures; ×7, PMO S1334. Bjørkåsholmen, Asker. Coll.: L. Størmer. ☐C. Dorsal view of right free cheek, showing facial suture; ×4, PMO 20061/2. Engervik, Asker. Coll.: W.C. Brøgger, 1879. ☐D. Dorsal view of hypostome retaining most of the exoskeleton; ×8, PMO 1402/2. Bjørkåsholmen, Asker. Coll. unknown, 1915. Figured by Skjeseth (1952, Pl. 4:14). ☐G, H. Lateral and dorsal view of hypostome, showing large anterior wings; ×5, PMO 121.564. Bjørkåsholmen, Asker. Coll.: F. Nikolaisen, 1960-05-01. ☐I. Dorsal view of moulting stage, showing very long pygidial spines; ×2.5, PMO 121.578/1. Bjørkåsholmen, Asker. Coll.: G. Henningsmoen, 1961-07-14. ☐J, L. Dorsal and lateral view of nearly complete specimen, showing flexibility of thoracic region; ×5.5, PMO 121.571/2. Bjørkåsholmen, Asker. Coll.: B.-D. Erdtmann, 1963. ☐K, M. Dorsal and lateral view of thoracic region with five segments; ×5, PMO 801/1. Stensberggaten, Oslo. Coll.: J. Kiær, 1911.

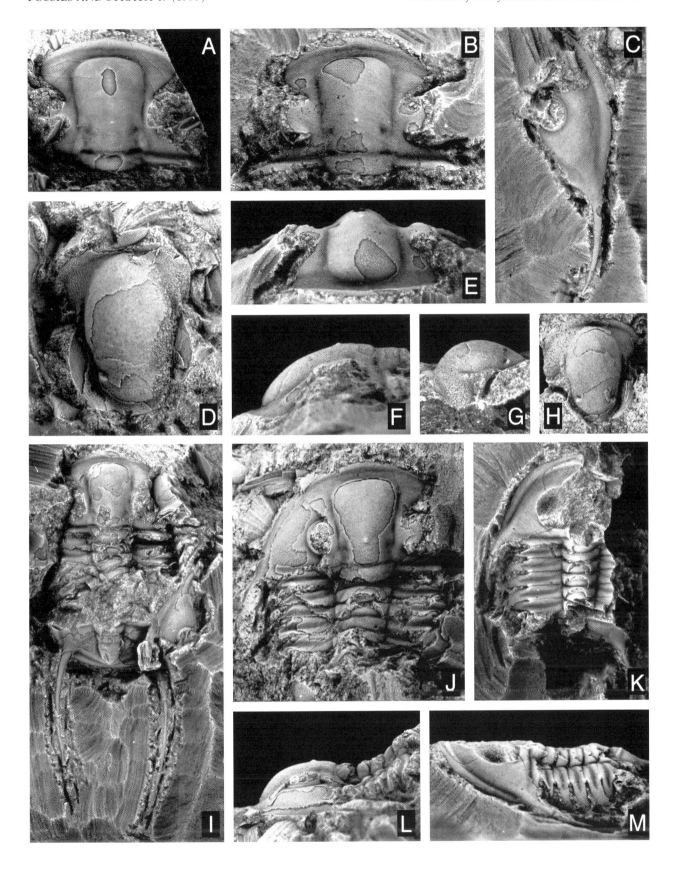

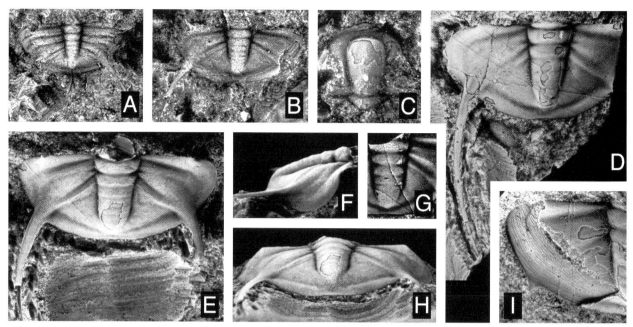

Fig. 59. Ceratopyge acicularis (Sars & Boeck, 1838). □A. Dorsal view of meraspid stage pygidium, showing initial thoracic segmentation; ×10, PM0 1366/1. Bjørkåsholmen, Asker. Coll. unknown, 1915. □B. Dorsal view of meraspid stage pygidium; ×10, PMO 1366/5. Bjørkåsholmen, Asker. Coll. unknown, 1915. □C. Dorsal view of meraspid stage cranidium with large glabellar tubercle; ×10, PMO 121.564/2. Bjørkåsholmen, Asker. Coll.: F. Nikolaisen, 1960-05-01. □D. Dorsal view of nearly complete pygidium; ×4, PMO S1348/1. Bjørkåsholmen, Asker. Coll.: L. Størmer. □E, F, H. Dorsal, lateral and posterior view of pygidium; ×5, PMO 70509/3. Bjørkåsholmen, Asker. Coll.: F. Nikolaisen, 1968-03-29. □G. Dorsal view of pygidium, showing exoskeletal granulation; ×4.5, PMO 121.658. Found in a dark limestone nodule 14–17 cm above base of the formation at Vestfossen railway st., Øvre Eiker. Coll.: J.O.R. Ebbestad, 1992-09-02. □I. Dorsal view of pygidium showing doublure; ×5, PMO 1334/3. Bjørkåsholmen, Asker. Coll. unknown, 1915.

489–489, Figs. 1–5. □1976b *Ceratopyge forficula acicularis* (Boeck) – Gjessing, pp. 134–141, Fig. 1A. □1989 *Ceratopyge forficula* (Boeck) – Bruton *et al.*, p. 237, Fig. 4:3, 6.

Lectotype. – Selected here; a partly preserved pygidium (PMO 57202a) from the old city cemetery in Oslo, Norway. This is one of only two pygidia in the collection at the Paleontological Museum in Oslo that positively were collected and identified by Boeck himself. Boeck (1838) described originally only the pygidium and emphasized the nature of the pygidial spines when distinguishing the new species. The lectotype specimen shows signs of having been prepared to reveal the pygidial spine. However, it cannot be established whether this was done by Boeck. The second specimen (PMO 57201) shows the nature of the rachis, which was also described by Boeck, and is selected as a paralectotype. Additional material of *C. acicularis* collected and labelled by Boeck includes two cranidia (PMO 56202b, c) and a free cheek (PMO 56202d). All specimens originate from the same locality.

Norwegian material and occurrence. – Exoskeletal parts are abundant throughout the formation across the Oslo Region and in the Lower Allochthon equivalent; the Solheim Slate, Ørnberget Formation at Groslii, Synfjell. Six almost complete specimens and six articulated cephala are also known. Tables 24 and 25 present cranidial and pygidial measurements, respectively, of the specimens figured herein.

Diagnosis. – Anterior width (tr.) of cranidium wider (tr.) than width (tr.) of palpebral lobes. Preglabellar field narrow (sag.). Pygidium with five rachial rings, adrachial flanks of pygidial spines extending from posterior margin posterior of rachial termination, not extending beyond anterior width of pygidium.

Emended description. – Cephalon semicircular, domed, with eyes situated high (Fig. 58L). Genal spines protruding backwards with an angle to margin. Sagittal

Table 24. Cranidial measurements of figured specimens of *Ceratopyge acicularis.*

Specimen	A	B	C	C1	C2	J	J1	J2	K	K1
S1334	0.60	0.53	0.24	0.15	0.12	0.87	0.59	0.57	0.29	0.30
93499	0.96	0.81	0.39	0.23	0.20	1.32	1.02	–	0.45	0.48
121.564/2	0.24	0.20	–	–	–	0.26	0.19	–	0.11	0.11
121.571/2	0.54	0.47	0.24	0.12	0.14	0.77	0.64	0.54	0.30	0.33
121.578/1	0.71	0.62	0.29	0.15	0.18	–	0.60	–	0.41	0.41

Table 25. Pygidial measurements of figured specimens of *Ceratopyge acicularis.*

Specimen	X	X1	Y	Z	W
S1348/1	0.32	0.18	0.65	0.72	1.44
1334/3	0.32	0.18	0.53	0.59	1.32
1366/1	0.05	0.03	0.14	0.15	0.31
1366/5	0.07	0.04	0.13	0.16	0.37
70509/3	0.23	0.18	0.45	0.51	1.14
121.658	0.27	0.14	0.45	0.50	1.08

length of cranidium two-thirds of posterior width. Glabella subrectangular, expanding slightly forward, making up 83% of sagittal length. Glabella slightly convex (tr.) medially, sloping steeply laterally (Fig. 58E) and to anterior glabellar field with almost transverse, bluntly truncated front (Fig. 58F). Occipital ring wide (sag.), making up one-third of posterior width, curving slightly crescent-like backwards, defined anteriorly by transverse occipital furrow, having deeper pits laterally than *C. forficula* (Fig. 58A, B). Glabella expands into faint bacculae just anterior of occipital ring, with 1S furrows as lateral pits or depressions situated opposite posterior extremities of palpebral lobes (Fig. 58A). The 2S and 3S furrows are not distinguishable. Median glabellar lobe positioned just anterior of 1S furrows, dorsal furrow but vaguely indicated. Posterior part of fixed cheek wide (tr.), spindle-shaped, curving only slightly laterally, with nearly transverse posterior margin. Posterior border furrow transverse, shallow, but distinct. Palpebral lobes tongue-shaped, short (sag.), width (tr.) at eyes equalling sagittal length. Anterior extremities of palpebral lobes situated just anteriorly to middle of glabella. Anterior facial suture diverging at an angle of 60° to the exsagittal line, extending outside exsagittal line of eyes before turning sharply inwards just posterior of anterolateral corners of glabella, to merge with semielliptical anterior margin. Anterior glabellar field distinct, with indistinct border furrow defining almost horizontal, bevelled border (Fig. 58F). Border attenuated laterally. Test on glabella with small granule.

Free cheeks elongated (Fig. 58C). Eye socle band-like, genal fields convex (tr.), curving down to narrow (tr.) border. At genal angle, genal field continues into long, thin genal spine, distinctly curved, not as continuation of semicircular cephalic margin. Suture opisthoparian.

Hypostome rectangular, median body large, domed, with no distinction of anterior or posterior lobes. Maculae small, situated far back. Posterior border almost effaced, lateral borders narrow (tr.), semielliptical, nearly merging with median body at its centre. Anterior wings wide (tr.), tongue-shaped (Fig. 58G). Anterior margin semielliptical with bevelled border (Fig. 58H). Test of median body with small pits, borders and anterior wings with terrace ridges (Fig. 58D).

Thorax with six segments. Rachis narrow (tr.), sides parallel. Pleurae horizontal, curving almost vertically laterally. Pleural furrows prominent, transverse, lying between low anterior and posterior flanges constricted laterally at transition to pointed outwards-directed ends of pleurae.

Sagittal length of pygidium slightly less than half posterior width. Rachis narrow (tr.) and long (sag.), tapering slightly backwards to bluntly rounded rachial termination, curving steeply down to lateral border. Rachis with five (Fig. 59D, E) rachial rings, anterior rachial ring well defined, posterior rings indicated by faint rachial furrows or lateral depressions only. Dorsal furrow distinct. Pleural field horizontal adrachially, curving steeply down to border laterally and posteriorly (Fig. 59F, H). Pair of wide (tr.) and deep pleural furrows extends obliquely backwards from anterior margins of rachis, attenuating laterally. Second and third pairs of furrows extend from dorsal furrow at posterior part of first rachial ring. Second pair distinct, short (sag.), transverse, almost effaced on the middle of the pleural fields, then outlining anterior flanks of pygidial spines laterally. Third pair of furrows not deep, diverging obliquely backwards at an angle of 55° to rachis, reaching to opposite posterior part of third rachial ring (counting from anterior margin). Posterior flanks of pygidial spines outlined by pair of faint depressions subparallel to latter pair of furrows, extending from posterior part of second rachial ring to border. Posterior border narrow sagittally, wider (sag.) close to pygidial spines. Posterior margin semicircular, slightly transverse posterior of rachis. Anterior margin transverse, defining raised anterior border (Fig. 59H). Long, slender pygidial spines extending backwards as coalescence of pleural field embraced by first and second rachial ring, gently curved, slightly upwards, inwards at posterior extremities, extending no wider than anterior width, being 3–2 times longer than sagittal length of pygidium (Figs. 58I and 59D, respectively). Their adrachial flank protrudes from margin distinctly anterior of rachial termination (Fig. 59D, E). Doublure not particularly wide, with fine terrace ridges (Fig. 59I). Test with small granules (Fig. 59G).

Ontogeny. – Several small specimens have been found. The smallest cranidium is 0.23 cm long (Fig. 59C), while the largest cranidum has a sagittal length of 1.41 cm, thus an increase in size of nearly 17 times. The smallest pygidium is 0.11 cm long (Fig. 59A).

On the meraspid cranidium the tubercle is larger, the glabella is narrower (tr.) with a large occipital ring, and the anterior width is smaller, with less divergent anterior facial sutures than in the holospis stage. The meraspid pygidium is very similar to the adult specimens, except that the pygidial spines are straighter, reaching beyond the anterior width of the pygidium (Fig. 59B). The adult proportions of the cranidium were established already when the sagittal length of the cranidium was 0.41 cm.

Discussion. – The Chinese species *C. transversa* Lu *in* Zhou *et al.*, 1977, differs from *C. acicularis* in having a larger tubercle, no preglabellar field, less divergent anterior facial suture, and a pygidium that is proportionally shorter (sag.), with only four rachial rings. In this respect it is similar to *C. forficula*.

C. elongata Chang & Fan, 1960, from China is very similar to *C. acicularis*, except that the glabella is proportionally longer and narrower (tr.) and the anterior facial sutures are much less diverging. The pygidium has only four rachial rings and more strongly defined pygidial spines.

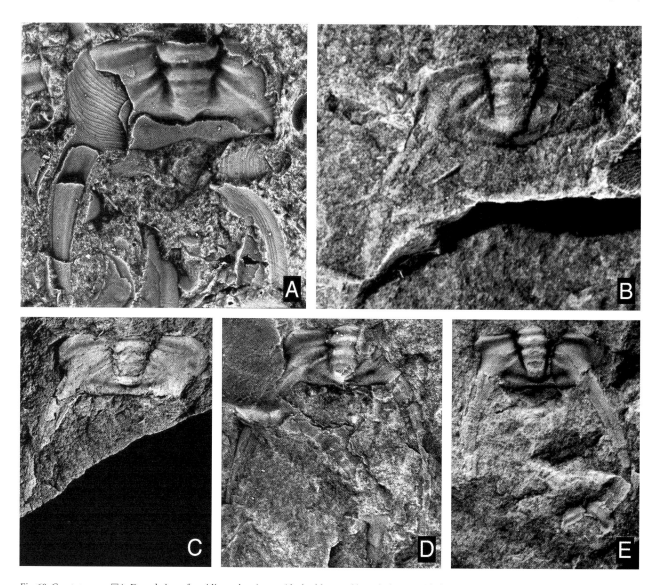

Fig. 60. Ceratopyge sp. □A. Dorsal view of pygidium, showing a wide doublure and broad, short pygidial spine; ×10, PMO 1272/1. Found in a dark lime-stone nodule 25–30 cm above base of the formation at Bjørkåsholmen, Asker. Coll. unknown, 1915. □B. Dorsal view of pygidium; ×13, PMO 56194h. Found in shale of the upper part of the Alum Shale Formation at Ruseløkkveien, Oslo. Coll.: M. Sars. □C. Dorsal view of large pygidium; ×6, PMO 56194i. Found in shale of the upper part of the Alum Shale Formation at Ruseløkkveien, Oslo. Coll.: M. Sars. □D. Dorsal view of pygidium with long pygidial spines; ×6.5, PMO 56193b. Found in shale of the upper part of the Alum Shale Formation at Ruseløkkveien, Oslo. Coll.: M. Sars. □E. Dorsal view of pygidium with preserved right lateral and posterior margin; ×8, PMO 56194e. Found in shale of the upper part of the Alum Shale Formation at Ruseløkkveien, Oslo. Coll.: M. Sars.

C. forficuloides Harrington & Leanza, 1957, from Argentina differs from *C. acicularis* mainly in having four pairs of lateral glabellar lobes, a tubercle closer to the occipital ring, a shorter (sag.) preglabellar field, and a wider (sag.) anterior border. *C. latilimbata* Moberg & Segerberg, 1906, from Sweden is distinguished by four pairs of lateral glabellar lobes, a very wide (sag.) preglabellar field, small palpebral lobes and indication of eye ridges (see Moberg & Segerberg, p. 87, Pl. 5:6). The latter species is only known from one specimen, and its affinity to *Ceratopyge* must be regarded with some caution.

Ceratopyge sp.

Figs. 55C, 56, 60

Material. – An external mould (PMO 1272/1), from a dark limestone nodule 25–30 cm above the base of the Bjørkåsholmen Formation at Bjørkåsholmen in Slemmestad, Norway. Four pygidia (PMO 56193b, d, i, PMO 56194e) from shale in the upper part of the Alum Shale Formation at Ruseløkkveien in Oslo, Norway. Table 26 presents measurements of the available pygidia.

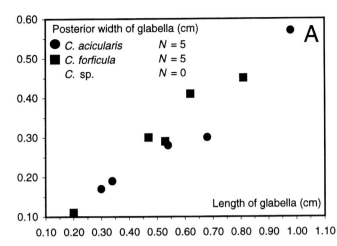
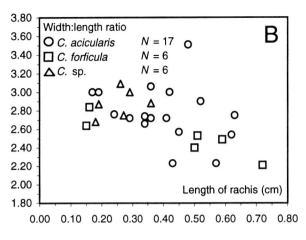

Fig. 61. Cranidia and pygidia of *Ceratopyge* species in the Alum Shale and Bjørkåsholmen Formations. □A. Length (sag.) of glabella plotted against posterior width (tr.) of glabella. □B. Length of rachis plotted against width:length of rachis ratio.

Occurrence. – The species is found in the dark limestone nodules near the base of the Bjørkåsholmen Formation and in shale from the upper part of the Alum Shale Formation

Description. – The pygidium of this species differs from *C. forficula* and *C. acicularis* in its subrectangular outline, proportionally shorter (sag.) than wide (tr.), rachis tapering more strongly posteriorly with only three rachial rings, all pleural furrows being wider (tr.), posterior margin curving anteriorly sagittally, pygidial spines having a broader base, being wider (tr.) generally two to nearly three times longer than sagittal length of pygidium, adrachial flank protruding from margin distinctly posterior of rachial termination. The doublure is wide (tr.) with open-spaced terrace ridges (Fig. 60A). *C.* sp. is possibly also smaller.

Discussion. – This species of *Ceratopyge* has probably contributed to the original description of *Ceratopyge forficula* Sars, 1835 (the specimens from the Alum Shale Formation were collected by Sars), but it represents a different form. However, awaiting more material, the species is left under open nomenclature.

The specimens found in the shale are somewhat different from those found in the limestone of the Bjørkåsholmen Formation. The distinct anterior curving posterior margin is less well developed, and there is a greater variety in the morphology of the genal spines. Some of these features may be related to the state of preservation, but it is likely that these specimens represent a variety of those found in the Bjørkåsholmen Formation. Searching localities for additional material in order to resolve this problem has not been possible, and in view of the great resemblance, they are here considered conspecific.

Fig. 61A, B presents different variables of cranidia and pygidia, respectively, of *Ceratopyge* species described herein.

Family Dikelokephalinidae Kobayashi, 1936

Genus *Dikelokephalina* Brøgger, 1896

Synonymy. – □1977 *Ciliocephalus* Liu – Liu *in* Zhou *et al.*, p. 227, Pl. 69:8, 9.

Type species. – *Centropleura? dicraeura* Angelin, 1854, p. 88, Pl. 41:9, from the Bjørkåsholmen Formation (upper Tremadoc) in Oslo, Norway; subsequently designated by Vodges (1925, p. 98).

Discussion. – Kobayashi (1936) and Lu (1975) listed the species belonging to this genus. Jell & Stait (1985) presented a diagnosis of the genus which is followed here, except that the pygidium of the type species has a high convexity, while the cranidium has a low convexity.

Several genera within the family share characters with *Dikelokephalina*. *Asaphopsis* Mansuy, 1920, which possesses pygidial spines at the posterolateral corners,

Table 26. Pygidial measurements of *Ceratopyge* sp..

Specimen	X	X1	Y	Z	W
1272/1	0.18	0.12	0.26	0.36	0.75
56193b	0.15	0.11	0.20	0.27	0.55
56193d	0.10	0.08	0.18	0.29	0.54
56194e	0.11	0.08	0.16	0.19	0.46
56194h	0.11	0.07	0.16	0.18	0.43
56194i	0.18	0.13	0.22	0.26	0.68

was regarded as closely related to *Dikelokephalina* by Kobayashi (1936) and Henningsmoen (1959), the former author pointing out that they might be members of a continuing series of gradual transitions. *Asaphopsides* Hupé, 1955, shares some *Asaphopsis* characters. Lu (1975) noted that the pygidia belonging to species of the Family *Dactylocephalus* Hsü *in* Hsü & Ma, 1948, have pygidial spines similar to those of *Dikelokephalina dicraeura* (Angelin, 1854), and a cranidium not unlike that of *Dikelokephalina asaiatica* Kobayashi, 1934. The type species of the genus *Ciliocephalus* Liu *in* Zhou *et al.*, 1977, discussed by Peng (1990), has a very strong similarity towards the type species of *Dikelokephalina*, separated only by minor morphological details.

The Dikelokephalinidae and the Taihungshaniidae were earlier associated, but Fortey (1981, p. 607) suggested that the latter may belong to the Cyclopygacea. Jell & Stait (1985) found that the Family Dikelocephalidae was most likely to contain the ancestral stock of *Dikelokephalina*, and the classification of the Asaphina by Fortey & Chatterton (1988) showed the latter suggestion to be the most probable.

Dikelokephalina dicraeura (Angelin, 1854)
Fig. 62

Synonymy. – □1854 *Centropleura? dicræura* n.sp. – Angelin, p. 88, Pl. 41:9. □1869 *Dikelocephalus dicræurus* (Angelin) – Linnarsson, p. 71. □1882 *Dicelocephalus dicræurus* (Angelin) – Brøgger, p. 126. □1896 *Dikelokephalina dicræura* (Angelin) – Brøgger, pp. 177–179, Text-fig. 4. □1900 *Dicellocephalina dicræura* (Angelin) – Moberg, p. 534, Pl. 14:1. □1906 *Dicellocephalina dicræura* (Angelin) – Moberg & Segerberg, p. 90, Pl. 5:12–14. □1919 *Dikelocephalina dicræura* (Angelin) – Lake, pp. 117–120. □1925 *Dikelokephalina dicraeura* (Angelin) – Vodges, p. 98. □1956a *Dikelokephalina dicraeura* (Angelin) – Tjernvik, p. 278. □1959 *Dikelokephalina dicraeura* (Angelin) – Harrington *et al. in* Moore, p. O359, Fig. 268:3a, b. □1959 *Dikelokephalina dicraeura* (Angelin) – Henningsmoen, p. 163, Pl. 1:1–4. □1961 *Dikelokephalina dicraeura* (Angelin) – Balashova, p. 125, Pl. 3:11–14. □1975 *Dikelokephalina dicraeura* (Angelin) – Lu, p. 155 (p. 358, English version). □1985 *Dikelokephalina dicraeura* (Angelin) – Jell & Stait, pp. 15–16. □1989 *Dikelokephalina dicraeura* (Angelin) – Bruton *et al.*, p. 235, Fig. 4:1.

Holotype. – A pygidium from the Bjørkåsholmen Formation in Oslo, Norway. Illustrated by Angelin (1854, Pl. 41:9). By monotypy. The original has not been identified in available collections in the Swedish Museum of Natural History, Stockholm, and until it is found or a neotype is selected, the Norwegian material described here must suffice to define the taxon.

Norwegian material. Two cranidia and six partly fragmentary pygidia. Found in the Bjørkåsholmen Formation in Slemmestad and Oslo in the Oslo Region, Aust Torpa, Snertingdalen, and Groslii, Synfjell, from the Lower Allochthon strata.

Remarks. – The cranidium of this species has been described earlier by Moberg (1900) and the pygidium by Henningsmoen (1959). However, the description of the cranidium is inadequate to fully cover the aspects of this species. Therefore, an emended description is provided here.

Emended description. – Posterior width of cranidium twice sagittal length. Glabella short (sag.) and narrow (tr.), making up half of sagittal length and one-fourth of posterior width, tapering forward, maximum width (tr.) of frontal glabellar lobe two-thirds occipital ring width (tr.). Occipital ring transverse with faint median occipital tubercle. Occipital furrow distinct laterally, faint medially. The 1S furrows are pit-like depressions, being forked adrachially, situated opposite middle (sag.) of glabella, halfway between sagittal line and dorsal furrow. The 2S furrows are deep, pit-like, situated opposite (exsag.) the 1S pits. The 3S furrows are pit-like, at the same line (exsag.) as the 1S and 2S pits. The 4S furrows are pit-like, situated marginally just anterolateral of the 3S pits. Faint marginal depressions present midway between 2S and 3S pits. Dorsal furrow faintly indicated. Posterior part of fixed cheeks wide (tr.) and narrow (sag.), making up one-fifth of sagittal length, lateral parts slightly wider. Posterior facial suture parallel to nearly transverse posterior margin, orientated exsagittally laterally. Border furrow centred on posterior limbs, without contact with glabella or lateral margin. Pair of alae present at fixed cheeks close to glabella, between occipital furrow and 1S pits. Fixed cheeks opposite palpebral lobes wide (tr.), palpebral lobes small, making up one-seventh of sagittal length, situated opposite middle of glabella, raised above fixed cheeks, weak indication of palpebral lobe furrows. Eye ridges extending obliquely forward from anterior extremity of eyes to 3S pits, continuing to outline anterolateral corners of glabella. Preglabellar field wide (sag. and tr.) making up half of sagittal length and four-seventh of posterior width. Subtriangular depression present between eye ridges and nearly transverse line in front of glabella. Anterior facial suture diverging with an angle of 45° to glabella, changing curvature just in front of glabella, merging with semielliptical anterior margin. Doublure wide (sag.) with open-spaced terrace ridges (Fig. 62A). Test mostly covered with small wrinkled terrace ridges.

The pygidium of this species was described by Henningsmoen (1959), and a redescription is not necessary

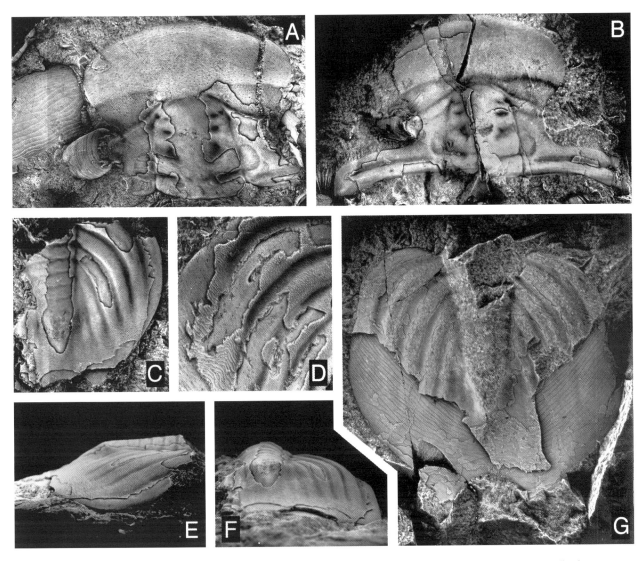

Fig. 62. Dikelokephalina dicraeura (Angelin, 1854). □A. Latex replica from external mould of incomplete cranidium, showing glabellar furrows; ×1.5, PMO 93893. Bjørkåsholmen, Asker. Coll.: Excursion 1973 (Trilobite symposium). □B. Dorsal view of cranidium, left part is a latex replica from external mould, showing palpebral lobe and facial suture; ×1.5, PMO 70248a. Bjørkåsholmen, Asker. Coll.: G. Henningsmoen, F. Nikolaisen, R. Ross & D.L. Bruton, 1968-06-12. □C, E, F. Dorsal, lateral and posterior view of pygidium, showing lateral margin and convexity of pygidium; ×2, PMO 847. Trefoldighetskirken, Oslo. Coll.: A. Rekdal, 1922. Figured by Henningsmoen (1959, Pl. 1:4). □D. Detail of exoskeletal surface of pleural field; ×3, PMO 83824. Bjørkåsholmen, Asker. Coll.: F. Nikolaisen, 1968-09-03. □G. Dorsal view of nearly complete pygidium, showing doublure and pygidial spines; ×2, PMO 69573. Roadcut west of Slemmestad, Røyken. Coll.: L. Størmer, 1953-10-23. Figured by Henningsmoen (1959, Pl. 1:2).

here. It may be added that there is a small knob at the posterior end of rachis, the postrachial ridge slopes steeply to the slightly concave border (Fig. 62E), the pleural field is horizontal adrachially, curving steeply down laterally (Fig. 62F) with six distinct pleural furrows curving backwards subparallel to margin, becoming progressively fainter backwards (Fig. 62C), test is covered with fine, small and wrinkled terrace ridges similar to those on the cranidium (Fig. 62D).

Free cheeks, hypostome and thoracic segments unknown.

Discussion. – Ciliocephalus angulatus Liu *in* Zhou *et al.*, 1977, is strikingly similar to *Dikelokephalina dicraeura*. Peng (1990) distinguished *Ciliocephalus* by a broadly conical, largely effaced glabella, proportionally shorter (sag.) preglabellar area, relatively broader (tr.) pygidium with a proportionally markedly shorter rachis less than half pygidial length (sag.), and shorter spines more openly spaced with the posterior margin gently curved anteriorly. The cranidia of the two species are, however, very similar. The preglabellar area of *D. dicraeura* is proportionally equal to that of *C. angulatus*, it has the same depressions in front of the eye ridges, and the proportions of the eyes and posterior fixed cheeks are almost identical. The rachis of *C. angulatus* is indeed proportionally shorter, but its length is seen to vary on the photographed material (Peng 1990, Pl. 17:2, 5, 8, 9). The outline and pygidial spines are comparable to those of *D. rugosa* Lu, 1975.

Concerning the strong resemblance between the two genera and the fact that *Ciliocephalus* has only two species assigned to it, its generic value is questioned. It seems better to treat *Ciliocephalus* as a synonym of *Dikelokephalina*.

Lake (1919) and Henningsmoen (1959) noted the great resemblance between *D. furca* (Salter, 1866) and the type species. They were suspected to be conspecific, but the material was too poor to decide whether this was true. Following the example of Hughes & Rushton (1990) with computer-aided restoration of tectonically distorted trilobites from Wales, the specimens of *D. furca* figured by Lake (1919, Pl. 14:11, 12) were restored for comparison between the two species. Lake's drawings were scanned and imported to the graphics program CorelDraw 4.0 (®Corel System Corporation). The images were manipulated to achieve a simple approximation to bilateral symmetry. The degree of compaction was not possible to estimate, and there is therefore still great uncertainty concerning the general shape of the British specimens. For that reason they are not figured here. Nevertheless, the restoration strengthens the overall similarity between the two species. Some differences exist, however. The posterior furrow in *D. furca* is in contact with both the glabella and the lateral fields, while it is centred (tr.) on the posterior fixed cheeks in *D. dicraeura*. The pygidium of the latter has six distinct rachial rings and a seventh indistinct ring. The pleural fields carry six distinct pleural furrows and a seventh indistinct one next to the terminal rachial piece. In *D. furca* there are no more than five distinct rachial rings, with a possible sixth indistinct ring. The pleural fields carry five distinct pleural furrows and a possible sixth indistinct furrow close to the terminal rachial piece. The features described above are distinct enough to warrant a separation between the two species.

Family Nileidae Angelin, 1854

Genus *Nileus* Dalman, 1827

Type species. – *Asaphus* (*Nileus*) *armadillo* Dalman, 1827, pp. 61–63 [246–248], 91 [276], Pl. 4:3, from Arenig limestone beds at Husbyfjöl in Östergötland, Sweden; by original designation of Dalman (1827, p. 61 [246]).

Discussion. – A diagnosis of the genus and a list of species of *Nileus* was given by Fortey (1975, p. 40). More than thirty species and subspecies are known from late Trema-

doc to Caradoc (Schrank 1972), with *Nileus limbatus* from Scandinavia being the oldest species. Nielsen (1995) provided an extensive description and discussion of this genus, with special emphasis on the phylogenetic relationship of Scandinavian species.

Based on the pygidial terrace-line patterns, three basic types of nileids were distinguished: an *exarmatus* type with almost smooth test surface; an *orbiculatoides* type similar to the previous, but with widely spaced terrace lines on pleural fields and sometimes on the entire border; and finally a *depressus* type with densely spaced terrace lines covering entire pygidium.

A strong facies dependency has been shown for the genus by Fortey (1975, 1980) in Spitsbergen and Nielsen (1995) in Scandinavia. *Nileus* is generally related to a muddy substrate. Nielsen (1995, p. 67) pointed out that the prominent terrace lines are strongly dependent on the environment and varies with depth.

Nileus limbatus Brøgger, 1882

Figs. 63, 64

Synonymy. – □1882 *Nileus limbatus* n.sp. – Brøgger, p. 62, Pl. 12:7. □1906 *Nileus armadillo* Dalman – Moberg & Segerberg, p. 93, Pl. 6:1–5. □1956a *Nileus limbatus* Brøgger – Tjernvik, p. 208, Text-fig. 33A, Pl. 2:12–15. □1961 *Nileus limbatus* Brøgger – Balashova, p. 131, Pl. 4:6. □1972 *Nileus limbatus* Brøgger – Schrank, pp. 356–357, Pl. 1:1–4. □1975 *Nileus limbatus* Brøgger – Fortey, p. 40. □1980 *Nileus limbatus* Brøgger – Tjernvik *in* Tjernvik & Johansson, p. 202. □1995 *Nileus limbatus* Brøgger – Nielsen, pp. 198–200.

Lectotype. – A pygidium (PMO H2710) from a dark limestone nodule near the base of the Bjørkåsholmen Formation at Vestfossen in Øvre Eiker, Norway. Illustrated by Brøgger (1882, Pl. 12:7a). Selected by Schrank (1972, p. 356).

Norwegian material. – Five cranidia, one free cheek, one hypostome, 11 pygidia. Tables 27 and 28 show the measurements of cranidia and pygidia, respectively.

Fig. 63. Nileus limbatus Brøgger, 1882. □A. Dorsal view of paralectotype cranidium, retaining most of the exoskeleton; ×10, PMO 1471. Found in a dark limestone nodule at Lunde in Vestfossen, Øvre Eiker. Coll.: W.C. Brøgger, 1879. Figured by Brøgger (1882, Pl. 12:7). □B, D, E. Dorsal, anterior and ventral view of cranidium (internal mould); ×10, PMO 973. Bjørkåsholmen, Asker. Coll.: V. Gaertner & T. Strand, 1928-05-13. □C, F. Dorsal and lateral view of right free cheek, showing terrace ridges and compound eye; ×10, PMO 966. Bjørkåsholmen, Asker. Coll.: V. Gaertner & T. Strand, 1928-05-13. □G, H, I. Dorsal, lateral and anterior

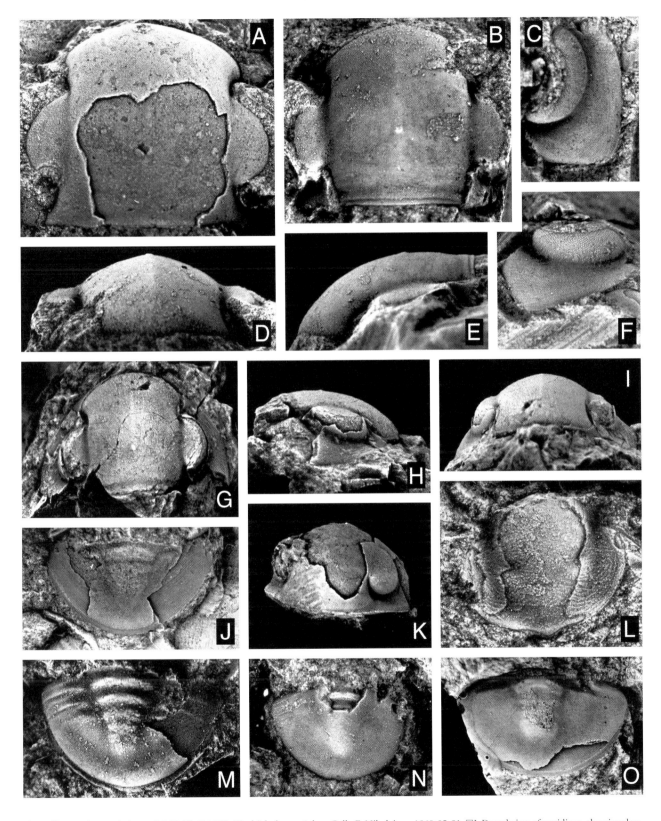

view of incomplete cephalon; ×5.5, PMO 121.593. Bjørkåsholmen, Asker. Coll.: F. Nikolaisen, 1960-05-01. □J. Dorsal view of pygidium, showing doublure; ×5, PMO 64121. Kutangen, Røyken. Coll.: L. Størmer, 1942-09-13. □K. Anterior view of incomplete cephalon, showing well-preserved eye; ×5.5,
PMO 136.085. Bjørkåsholmen, Asker. Coll.: L. Størmer, 1955-04-29. □L. Dorsal view of hypostome; ×12, PMO 83940/5. Road section west of Slemmestad, Røyken. Coll.: G. Henningsmoen, 1950. □M. Dorsal view of meraspid stage pygidium, showing initiation of thoracic segmentation (internal
mould); ×8, PMO 888/1. Sofienberg, Oslo. Coll.: J. Kiær, 1910. □N. Dorsal view of meraspid stage pygidium with preserved exoskeleton; ×8, PMO
121.619. Locality unknown. Coll.: F. Nikolaisen. □O. Dorsal view of short (tr.) pygidium; ×8, PMO 1111. Engervik, Asker. Coll.: W.C. Brøgger, 1879.

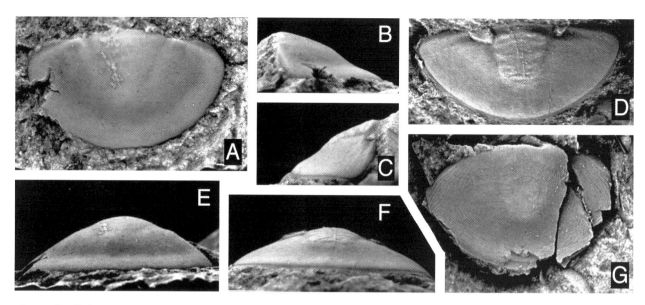

Fig. 64. Nileus limbatus Brøgger, 1882. □A, B, E. Dorsal, ventral and posterior view of lectotype pygidium; ×8, PMO H2710. Found in a dark limestone nodule at Vestfossen, Øvre Eiker. Coll.: W.C. Brøgger, 1879. Figured by Brøgger (1882, Pl. 12:7a). □C, D, F. Lateral, dorsal and posterior view of pygidium (internal mould); ×8, PMO 121.630. Bjørkåsholmen, Asker. Coll.: F. Nikolaisen, 1960-08-05. □G. Dorsal view of pygidium, showing distinct terrace ridges; ×10, PMO 64117/3. Kutangen, Røyken. Coll.: L. Størmer, 1942-09-13.

Remarks. – A diagnosis was provided by Schrank (1972, pp. 356–357). A full description emended from Brøgger (1882) and Moberg & Segerberg (1906) is provided here.

Emended description. – Cephalon semielliptical in outline, very convex (tr.), curving steeply down laterally (Fig. 63I). Anterior arch wide (tr.).

Sagittal length of cranidium slightly less than posterior width. Glabella convex (tr.), evenly curved abrachially, forming a faint sagittal line (Fig. 63G). It occupies the entire width (tr.) and length (sag.) of cranidium, slanting down anteriorly (Fig. 63E), parallel-sided between palpebral lobes, expanding laterally in front of eyes. Occipital ring indistinct with faint, transverse occipital furrow (Fig. 63A, B, E; not including the posteriormost ring in the two latter). Median occipital tubercle centred between eyes (Fig. 63B, E). Posterior fixed cheeks short (tr.), triangular, merged with glabella. Posterior margin transverse. Palpebral lobes large, semielliptical, making up slightly more than two-fifths of sagittal length, situated close to and opposite the posterior two-thirds of glabella. Sagittal length 80% of width (tr.) at eyes. Front of anterior glabellar lobe evenly rounded.

Free cheeks small, subtriangular. Genal field smooth with terrace ridges (Fig. 63C, F). Compound eyes prominent (Fig. 63C, F, G–I, K).

Hypostome: One specimen (Fig. 63L) resembling that figured by Tjernvik (1956a, Pl. 2:12), wider (tr.) than long (sag.). Anterior lobe of median body occupying half sagittal length. Indistinct transition to posterior lobe, indicated by pair of faint lateral maculae. Lateral border wide (tr.), merging with median body opposite middle of anterior lobe. Margin semicircular, slightly pointed sagittally.

Sagittal length of pygidium between 45% and 60% of anterior width (Figs. 64D and 63O, respectively). Rachis short (sag.) and wide (tr.), tapering markedly posteriorly of anterior margin, then nearly parallel-sided backwards to the almost transverse posterior rachial termination (Figs. 63O, 64D). Faint depressions indicating

Table 27. Cranidial measurements of *Nileus limbatus.*

Specimen	A	B	C	C1	C2	J	J1	J2	K	K1
S1150	–	–	0.15	0.26	–	–	0.54	0.78	–	0.54
973	0.55	0.55	0.14	0.17	0.06	0.60	0.38	0.60	0.33	0.38
1471	0.56	0.56	0.18	0.24	0.11	0.65	0.74	0.52	–	1.05
121.593	0.60	0.60	0.21	0.26	–	–	0.49	0.71	0.48	0.49
136.085	–	–	0.13	0.18	–	0.42	0.54	–	0.42	–

Table 28. Pygidial measurements of *Nileus limbatus.*

Specimen	X	X1	Y	Z	W
H2710	0.27	0.10	0.26	0.38	0.72
S3033/2	0.15	0.08	0.21	0.31	0.54
888/1	0.15	0.09	0.21	0.30	0.44
1111	0.19	0.11	0.26	0.39	0.70
1272/2	0.13	–	0.14	0.18	0.43
64117/3	0.19	0.09	0.24	0.37	0.62
64121	0.27	0.15	0.33	0.49	0.84
83968/3	0.17	0.07	0.17	0.23	0.39
121.590	0.21	0.11	0.27	0.40	0.70
121.619	0.18	0.10	0.19	0.26	0.41
121.630	0.21	0.12	0.20	0.33	0.72

segmentation, visible on internal moulds (Fig. 63M). Relief of rachis low, with smooth transition to evenly curved (tr.), slightly convex pleural fields (Fig. 64E). Concave transition (sag.) to nearly flat posterior margin (Fig. 64B, C). Pygidial border wide, not well defined on internal moulds. Terrace ridges at margin. Outline of posterior margin semicircular, anterior margin nearly transverse, with large (tr.) articulating facets (Fig. 64D). Doublure wide (tr.), with densely spaced terrace ridges (Fig. 64J).

Discussion. – This is the oldest known species of *Nileus*. The lectotype and paratype specimens come from the dark limestone nodules at the base of the Bjørkåsholmen Formation, but the species occurs frequently in the light-grey limestone of the same formation. Fig. 65 shows dimensions of the pygidia, with indication of stratigraphical occurrence.

The only apparent differences between the specimens are the better defined pygidial border of the lectotype (Fig. 64A), which is comparable to that of the early Arenig *N. exermatus* Tjernvik, 1956, from Sweden, and while the exoskeleton of the holotype pygidium shows but faint, marginal terrace ridges, the single pygidium preserved with an exoskeleton from the light-grey limestone has fine terrace ridges covering the entire test (Fig. 64G). They are subparallel to the margin laterally, transverse anteriorly and adrachially, and appear similar in structure to those of the early Arenig species *N. glazialis* Schrank, 1973 from N. E. Germany, and the late Arenig subspecies *N. glazialis costatus* Fortey, 1975 from Spitsbergen. The distinction of the terrace ridges is better in these two younger species.

Several distinct *Nileus* populations with several subspecies have been reported from the early Ordovician of Sweden (Fortey 1975, p. 41, Nielsen 1995). Nielsen (1995, pp. 197–201) distinguished three systematically important basic types based on the pygidial terrace-line patterns: *N. exarmatus* type with a smooth pygidium with marginal terrace lines only; *N. orbiculatoides* type similar to the former type, but with widely spaced terrace lines across the pleural fields and occasionally on the entire border; *N. depressus* type with densely spaced terrace lines across the pygidium. The available material in this study is too limited to allow definitive conclusions of the variation within the Bjørkåsholmen Formation. Of the eleven pygidia known, only two are from the dark limestone nodules at the base of the formation. One of the paratype pygidia shows marginal terrace lines, placing it in the *N. exarmatus* type (Fig. 64A). The remaining specimens are mostly internal moulds found in the overlying light-grey limestone. Only one specimen has shell preserved, showing widely spaced terrace lines of the *N. orbiculatoides* type (Fig. 64G). These differences would warrant a separation, following the concept of basic types of Nielsen (1995, p.

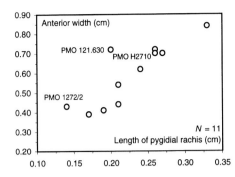

Fig. 65. Pygidia of *Nileus limbatus*. Length (sag.) of pygidial rachis plotted against anterior width (tr.). PMO H2710 (lectotype) and PMO 1272/2 come from black limestone nodules at the base of the formation. PMO 121.630 (Fig. 63D) has an exceptionally short pygidial rachis.

197). However, since the remaining specimens in the light-grey limestone are internal moulds, assigning these to either of the two basic *Nileus* types is unwise. The available cranidia do not tell much more. The cranidia found in the dark limestone nodules appear to have a somewhat higher profile, slightly stronger convexity, the tubercle perhaps placed more anteriorly, and no median line anterior of the tubercle. These features may be within the acceptable variations. but a larger collection is needed to evaluate possible differences.

From the light-grey limestone, one pygidium (Fig. 64D) is distinguished by being proportionally wider (tr.) and having a slightly more convex (sag.) transition to the posterior border marginally (Fig. 64C).

Genus *Symphysurus* Goldfuss, 1843

Type species. – *Asaphus palpebrosus* Dalman, 1827, pp. 60–61 [245–246], 91 [276], Pl. 4:2, from Husbyfjöl in Västergötland, Sweden; subsequently designated by Barrande (1852, p. 654).

Discussion. – The type species was revised by Fortey (1986). *Symphysurus* is an exceptionally widespread genus, present at several palaeolatitudes in the early Ordovician, and is morphologically a rather coherent group. Fortey (1986, pp. 266–273) argued for an infaunal mode of life of this genus on the marginal parts of platforms.

Symphysurus angustatus (Sars & Boeck, 1838)

Figs. 66–68

Synonymy. – □1838 *Trilobites angustatus* n.sp., Sars & Boeck mscr. – Boeck, p. 142. □1869 *Symphysurus socialis* n.sp. – Linnarsson, p. 74, Pl. 2:33, 34. □1882 *Symphysurus angustatus* (Boeck) – Brøgger, p. 60, Pl. 3:9–11. □1902 *Symphysurus angustatus* (Boeck) – Pompeckj, pp. 3–4, Fig. 1. □1906 *Symphysurus angustatus* (Boeck) – Moberg & Segerberg, pp. 90–91, Pl. 5:15–21. □1940 *Symphysurus angustatus* (Boeck) – Størmer, p. 143. □1956a *Symphysurus* (*Symphysurus*) *angustatus* (Sars & Boeck) – Tjernvik, pp. 211–212, Pl. 2:24, 25. □?1973 *Symphysurus angustatus* (Boeck) – Modlinski, pp. 51–52, Pl. 3:5, 6. □1973 *Symphysurus angustatus* (Boeck) – Dean, p. 328. □*non* 1975 *Symphysurus angustatus* (Boeck) – Courtessole & Pillet, pp. 263–266, Pl. 26:1–19, Pl. 27:2, 3. □1986 *Symphysurus angustatus* (Boeck) – Fortey, p. 261. □*non* 1986 *Symphysurus angustatus angustatus* (Boeck) – Berard, Pl. 9:4, 6, 10.

Lectotype. – A cranidium (PMO 56215a) from the Bjørkåsholmen Formation in Oslo. Selected by Størmer (1940, p. 143).

Norwegian material. – Cranidia, pygidia and free cheeks occur abundantly in the formation across the Oslo Region. In addition, two nearly complete specimens, two cephala and three hypostomes are known.

Remarks. – *Symphysurus angustatus* has been widely referred to and partly diagnosed and described (Boeck 1838; Linnarsson 1869; Brøgger 1882). So far, however, a detailed diagnosis and full description have not been provided.

Emended diagnosis. – Glabella large, convex (tr.), expanding slightly forward over anterior margin, expanding posterior of palpebral lobes. Occipital furrows indicated marginally on glabella. Palpebral lobes semicircular, centred slightly posterior of the middle of the glabella. Thorax with eight segments. Pygidium semicircular, convex, posterior width slightly less than two times sagittal length. Rachis with four rachial rings, visible on internal moulds, making up three-fourths of sagittal length. Exoskeleton covered with fine terrace ridges.

Emended description. – Sagittal length of cranidium about three-fourths of or subequal to posterior width, depending on how measurements are taken. Glabella large, convex (tr.), occupying nearly two-thirds of posterior width and entire length (sag.) of cranidium, curving steeply down anteriorly, expanding but slightly past anterior margin (Fig. 66H). It expands (tr.) slightly anterior of eyes, rounded in front, parallel-sided backwards. Transverse profile high, slightly pointed medially, gently convex laterally, with abrupt, almost vertical transition to border furrow. Occipital furrow indicated by faint, lateral depressions on internal moulds. Posterior margin of glabella weakly crescentic backwards. Median glabellar tubercle positioned slightly posterior to the middle of glabella (Fig. 66G, H). Posterior part of fixed cheeks short (tr.), horizontal close to glabella, triangular in outline. Transverse border furrows indicated on internal moulds.

Palpebral lobes large, making up one-third of sagittal length, positioned slightly posterior to the middle of glabella. Margin semielliptical, curving inwards at posterior and anterior extremities. Anterior part of palpebral lobes positioned closer to glabella than posterior part, maximum width (tr.) at eyes being one-fourth wider than sagittal length.

Free cheeks small, posterior part of genal field subtriangular, anterior part narrow (tr.). Eye socle low, brim-like, outlined by furrow which becomes deeper anteriorly (Fig. 66B). Eyes large (Fig. 66B–E). Genal field with terrace ridges subparallel to smooth, well-rounded margin (Fig. 67B). Free cheeks merged with entire cephalic doublure (Fig. 66N). Doublure wide sagittally, with terrace ridges. Connection to hypostome indicated by triangular depression (Fig. 66M). Small vincular notches present at posterior extremities (Fig. 66M, N).

Hypostome subrectangular, wider than long. Median body occupying two-thirds of sagittal length, tapering slightly backwards, imitated by weakly indicated, lateral maculae. Posterior margin meeting low angle sagittally, lateral parts converging slightly anteriorly, not meeting median body (Fig. 66I). Anterior wings wide (tr.), positioned nearly vertical, making up half the anterior width (Fig. 66F). Test with terrace ridges.

Thorax with eight segments. Rachial rings narrow (sag.), with large articulating half-ring curving anteriorly. Rachial furrows deep, transverse. Pleural fields curving slightly backwards. Triangular pleural furrows indicated on internal moulds (Fig. 66J).

Sagittal length of pygidium slightly more than half posterior width. Rachis narrow (tr.) tapering posteriorly, making up three-fourths of sagittal length, curving steeply down to posterior margin (Fig. 67N). Four rachial rings indicated by faint lateral furrows on internal moulds. Pleural fields convex (tr.) curving steeply down laterally

Fig. 66. Symphysurus angustatus (Sars & Boeck, 1838). □A. Latex replica from external mould of nearly complete specimen; ×5.5, PMO 83794/1. Nærsnes, Røyken. Coll.: N. Spjeldnæs, 1952. □B. Lateral view of cephalon, showing anterior facial suture; ×2.5, PMO 121.585/1. Bjørkåsholmen, Asker. Coll.: R. Ross, 1968-06-12. □C, D, E. Dorsal, anterior and lateral view of partly complete cephalon, showing the large eye; ×3, PMO 136.083. Bjørkåsholmen, Asker. Coll.: B. Funke, 1987. □F. Hypostome with exoskeleton preserved; ×5, PMO H2688/2. Vestfossen, Øvre Eiker. Coll.: W.C. Brøgger, 1879. □G, H, K. Dorsal, lateral and anterior view of cranidium, showing terrace ridges; ×2.5, PMO 83993. Grundvik in Nærsnes, Røyken. Coll.: G. Henningsmoen, 1955. □I. Internal mould of hypostome; ×10, PMO 1301/2. Bjørkåsholmen, Asker. Coll. unknown, 1915. □J. Latex replica from external mould of thoracic segment; ×2.5, PMO 1297/3. Bjørkåsholmen, Asker. Coll. unknown, 1915. □L. Dorsal view of palpebral lobe, showing terrace ridges; ×9, PMO 1442/2. Vestfossen, Øvre Eiker. Coll.: W.C. Brøgger, 1879. □M. Ventral view of doublure; ×4.5, PMO S1341/3. Bjørkåsholmen, Asker. Coll.: L. Størmer, 1918? □N. Ventrolateral view of doublure and free cheek, showing lateral pits; ×9, PMO 1443. Vestfossen, Øvre Eiker. Coll.: W.C. Brøgger, 1879.

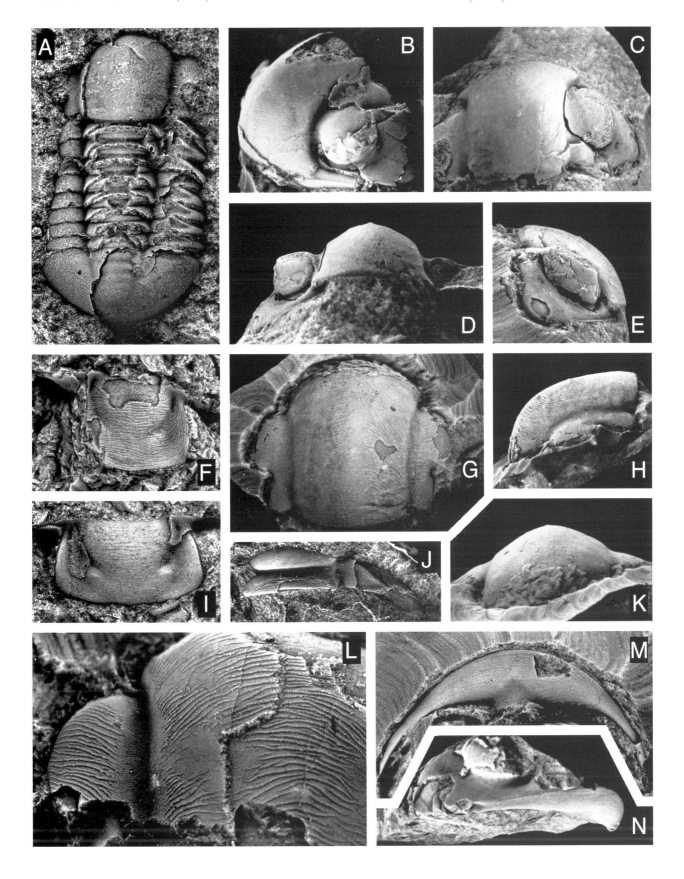

Genus *Varvia* Tjernvik, 1956

Type species. – *Symphysurus breviceps* Angelin, 1854, p. 61, Pl. 33:13, from the Lower Planilimbata Limestone (Arenig) at Oltorp in Västergötland, Sweden; by original designation of Tjernvik (1956a, p. 212).

Discussion. – Based on the distinct median suture, Tjernvik (1956a) assigned this genus to the Family Asaphidae. It was later (Harrington *et al. in* Moore 1959) reassigned to the Subfamily Symphysurininae, again distinguished by its median suture. However, Fortey (1983) found that this subfamily could not belong to within the Asaphidae, and subsequently the genus *Symphysurina* was considered the only representative of the subfamily (Fortey & Chatterton 1988). Though it has a well-developed median suture, *Varvia* is best placed within the Family Nileidae. Early representatives of the Nileidae had a median suture (Fortey & Chatterton 1988). Furthermore, *Varvia* appears to have seven or eight thoracic segments and a wide hypostome with broad borders (Tjernvik 1956). The many features shared with *Symphysurus*, for instance both being sub-isopygous and both having rounded genal angles, similar anteriorly positioned eyes, a similar rounded and expanded frontal part of the glabella, and similar wide hypostomes, also suggest a common family affinity. Courtessole & Pillet (1975) questionably included *Varvia* in the Subfamily Lakaspidinae, one of seven subfamilies of Nileidae (see Fortey & Chatteron 1988 for comments on the validity of these subfamilies).

The genus *Varvia* has possibly four species, *Varvia breviceps* (Angelin, 1854), *Varvia falensis* Tjernvik, 1956 (both lower Arenig), and *Varvia longicauda* Tjernvik, 1956 (upper Tremadoc), all from Baltoscandian strata, and *Varvia*? sp. *sensu* Ross (1970, p. 76) from the Llanvirn of Nevada, USA. The latter assignment is dubius. *Varvia* seems endemic to the Baltoscandia area, and the species serve as index fossils. In Norway *V. longicauda* is found in the Bjørkåsholmen Formation, and Hoel (in press) noted *V. breviceps* in the Biozone of *Megistaspis planilimbata* at Vestfossen in the Eiker–Sandsvær district.

Varvia longicauda Tjernvik, 1956

Fig. 69

Synonymy. – □1956a *Varvia longicauda* n.sp. – Tjernvik, p. 215, Text-fig. 34A, Pl. 3:10, 11.

Holotype. – An almost complete specimen (PMU VG 248) from the Bjørkåsholmen Formation at Stenbrottet in Västergötland, Sweden. Identified and figured by Tjernvik (1956a, p. 215, Pl. 3:11).

Norwegian material. – Numerous cranidia are found in the collection at the Paleontological Museum in Oslo, but only five pygidia are recognized.

Discussion. – The diagnosis and description of this species given by Tjernvik (1956a, pp. 215–216) is adequate and adopted here, except that the anterior margin of the pygidium is transverse near rachis, indicating narrow (tr.), horizontal pleural fields close to the rachis (Fig. 69D), and that the transition between the rachis and posterior border is first slightly concave, then convex marginally (Fig. 69G), a rather distinct feature of this species. None of the Norwegian specimens show any terrace ridges on the pleural fields, but at the anterolateral corners of the pygidium some small terrace ridges are present (Fig. 69I).

Cranidia of *Symphysurus angustatus* (Sars & Boeck, 1838) and *Varvia longicauda* are difficult to distinguish. In general, *S. angustatus* has the glabellar tubercle positioned opposite the middle (sag.) of the palpebral lobes, in contrast to near the posterior extremities in *V. longicauda*, less rounded glabellar front expanded more markedly anterolaterally, and the lateral transition to the border furrow is vertical. In *V. longicauda*, the glabella has a lower transverse profile, curving (tr.) gently down laterally to border furrow. The pygidium of this species is easily confused with that of *Nileus limbatus* Brøgger, 1882, and small specimens of *Symphysurus angustatus*. They can be distinguished by the transition between the rachis and posterior border, being first slightly concave and then distinctly convex marginally in *V. longicauda*, more prominently concave in *N. limbatus* and convex in *S. angustatus*. Some interference in these characters is possible owing to state of preservation.

Family Panderiidae? Bruton, 1968

Genus *Ottenbyaspis* Bruton, 1968

Type species. – *Illaenus oriens* Moberg & Segerberg, 1906, p. 98, Pl. 7:2a–c, from the Bjørkåsholmen Formation (upper Tremadoc), at Ottenby on Öland, Sweden. By original designation of Bruton (1968, p. 29).

Discussion. – The genus comprises four species, *Ottenbyaspis oriens* (Moberg & Segerberg, 1906) from the upper Tremadoc of Norway and Sweden, *O. perseverens* (Tjernvik, 1956) from the lower Arenig of Sweden, *O.* sp. *sensu* Poulsen (1965) and Nielsen (1995) from the upper Arenig of Bornholm, Danmark, and *O.? broeggeri* (Růžička, 1926) from the Tremadoc of central Bohemia, Czech Republic.

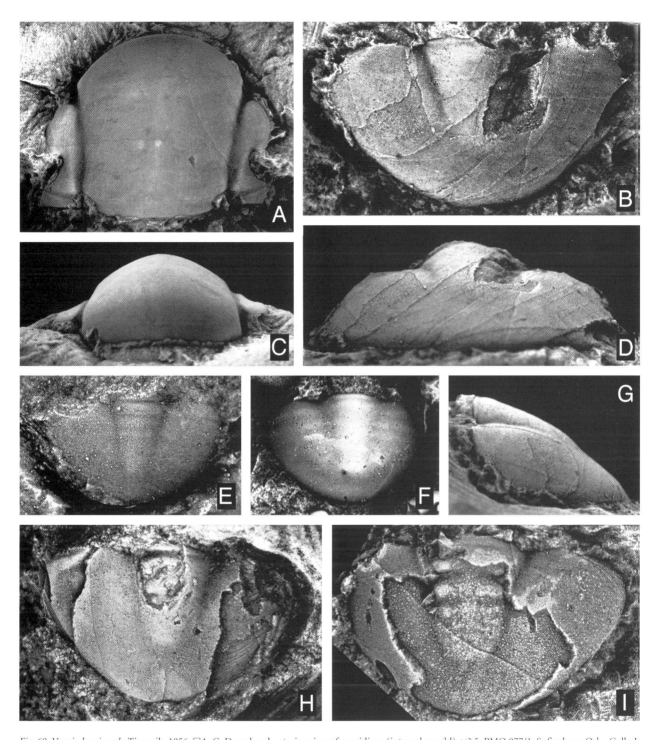

Fig. 69. Varvia longicauda Tjernvik, 1956. □A, C. Dorsal and anterior view of cranidium (internal mould); ×3.5, PMO 877/1. Sofienberg, Oslo. Coll.: J. Kiær, 1910-04-25. □B, D, G. Dorsal, posterior and lateral view of pygidium; ×8, PMO 832. Stensberggaten, Oslo. Coll.: J. Kiær. □E. Dorsal view of meraspid stage pygidium; ×14, PMO 121.581. Bjørkåsholmen, Asker. Coll.: G. Henningsmoen, 1959-02-22. □F. Dorsal view of meraspid pygidium; ×16, PMO 1202/5. Found in a dark limestone nodule 25–30 cm above the base of the formation in Slemmestad. Coll. unknown, 1915. □H. Dorsal view of pygidium retaining most of the exoskeleton; ×12, PMO 1412/3. Fure, Modum. Coll. unknown, 1891. □I. Dorsal view of pygidium, showing anterolateral terrace ridges; ×10, PMO 83994. Grundvik in Nærsnes, Røyken. Coll.: G. Henningsmoen, 1955-09-12.

Lane & Thomas (1983) followed Bruton (1968) in placing *Ottenbyaspis* in the Family Panderiidae Bruton, 1968, but expressed doubts since the nature of the ventral sutures are unknown for the species. The Norwegian material adds no new information to this, and the assignment to Panderiidae is not altered.

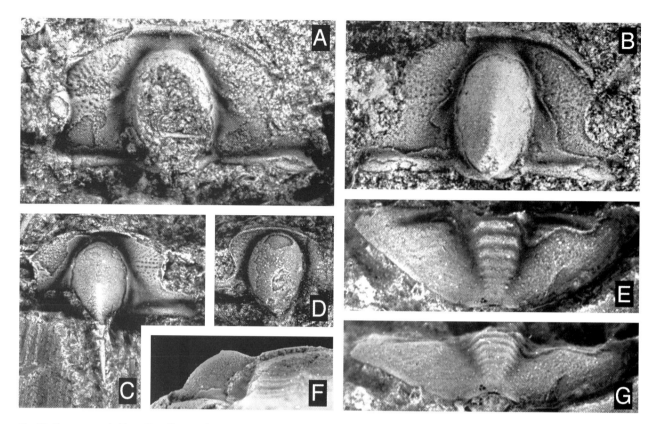

Fig. 72. Orometopus elatifrons (Angelin, 1854). □A. Dorsal view of incomplete cranidium, showing eye and eye ridge; ×12, PMO 1381/2. Kutangen, Røyken. Coll.: L. Størmer, 1942-09-13. □B. Dorsal view of incomplete cranidium, showing anterior and posterior margins; ×12, PMO 84066/2. Bjørkåsholmen, Asker. Coll.: G. Henningsmoen, 1958-11-16. □C, F. Dorsal and lateral view of complete cranidium; ×10, PMO 121.558/2. Bjørkåsholmen, Asker. Coll.: F. Nikolaisen, 1960-05-01. □D. Dorsal view of meraspid stage cranidium; ×12, PMO 136.022. Found 10–18 cm above base of the formation at Færdenveien in Klekken, Ringerike. Coll.: J.O.R. Ebbestad, 1992-09-09. □E, G. Dorsal and posterior view of pygidium; ×16, PMO 872/5. Rodeløkken, Oslo. Coll.: J. Kiær, 1910.

wards posteriorly, continuing backwards into long, thin spine (Fig. 72F). Posterior margin of glabella outlined by collar-like occipital ring raised above fixed cheeks (Fig. 72B). Anterior border distinct anteriorly, outlining anterior half of glabella. Three pairs of pit-like depressions at anterior border. Posterior pair between posterior margin and glabellar tubercle, placed on baccula-like area (Fig. 72C), second pair positioned where eye ridges meet glabella, third pair positioned at anterolateral corners of glabella. Posterior margin transverse. Posterior part of fixed cheeks wide (tr.), with narrow (sag.) lateral limbs making up 29% of sagittal length. Posterior border furrow distinct, wide (tr.), lateral termination pit-like. Palpebral lobes small, semicircular, making up one-fifth of sagittal length, situated marginally on wide (tr.), horizontal fixed cheek. Eye ridges converge anteriorly from just anterior of mid-length (sag.) of palpebral lobes (Fig. 72A). Anterior facial suture straight (exsag.) or slightly diverging, merging with subelliptical anterior margin just posterior of glabellar front. Preglabellar area short (sag.). Anterior border raised, wire-like, transverse sag-

ittally, curved backwards laterally, with terrace ridges. Distinct small pits around eye ridges.

Free cheeks, hypostome and thorax unknown.

Pygidium triangular, sagittal length one-third of anterior width (Fig. 72E). Rachis narrow (tr.), tapering backwards to margin, with seven distinct rachial rings. Posterior margin slightly transverse posterior of rachis, with distinct posterior arch (Fig. 72G). Furrow near anterior margin, short, deep. Anterior margin transverse.

Discussion. – Of the eleven species listed above, only three (e.g., *Orometopus elatifrons, O. grypos, O. klouceki*) possess a backward-projecting glabellar spine; *O. elatifrons* possesses the only glabella with an oval outline. The glabella of the other species either expands slightly anteriorly or markedly, as in *O. pyrifrons* and *O. pyrus*, or are subparallel-sided, as in *O. praenuntius* and *O. notatifrons*.

The subrectangular outline of the prepalpebral area of the cranidium is another typical feature of this species. *O. pyrus* and *O. pyrifrons* are distinguished by an anterior margin subparallel to the expanded front of the glabella, and by the apparent lack of a bevelled anterior border,

Table 29. Cranidial measurements of *Orometopus elatifrons.*

Specimen	A	B	C	C1	C2	J	J1	J2	K	P
S1152	0.24	0.20	–	–	–	0.47	0.25	–	0.12	–
S1168/2	0.18	0.15	–	–	–	–	0.20	–	0.11	–
S1169/2	0.22	0.19	0.09	0.05	0.06	0.40	0.28	–	0.12	–
S1201	0.21	0.18	–	–	–	0.48	0.30	–	0.12	–
S1202	0.16	0.15	–	–	–	–	0.26	–	0.09	–
S1203/2	0.16	0.15	–	–	–	–	0.26	–	0.10	–
S1205	0.30	0.25	–	–	–	–	0.32	–	0.18	–
325/8	0.22	0.18	–	–	–	0.48	0.30	–	0.12	–
774/1	0.21	0.18	0.08	0.05	0.06	0.47	0.30	–	0.12	–
790/1	0.21	0.19	–	–	–	–	0.28	–	0.12	–
797/1	0.25	0.22	0.10	0.06	0.06	–	0.38	–	0.14	–
950/1	0.24	0.22	0.10	0.04	0.06	0.52	0.31	0.38	0.13	0.14
962/3	0.18	0.15	–	–	–	–	0.26	–	0.12	–
964/5	0.16	0.15	–	–	–	–	0.28	–	0.10	–
1087/1	0.17	0.14	–	–	–	0.36	0.25	–	0.08	0.10
1088	0.15	0.13	–	–	–	0.38	0.24	–	0.08	–
1089	0.12	0.10	–	–	–	0.31	0.18	–	0.06	–
1156/3	0.20	0.18	–	–	–	0.44	0.28	–	0.12	–
1164/3	0.20	0.17	0.07	0.04	0.05	0.38	0.26	–	0.11	–
1286/6	0.18	0.15	–	–	–	0.38	0.23	–	0.09	–
1292/1	0.21	0.18	–	–	–	0.54	0.39	–	0.14	–
1299/2	0.15	0.13	–	–	–	0.32	0.24	–	0.08	–
1299/3	0.22	0.19	–	–	–	–	0.34	–	0.13	–
1349/3	0.27	0.22	–	–	–	–	0.32	–	0.13	–
1351/2	0.19	0.15	–	–	–	–	0.26	–	0.11	–
1376/3	0.18	0.16	–	–	–	–	0.28	–	0.11	–
1381/2	0.28	0.23	0.11	0.07	0.07	0.50	0.40	0.44	0.17	–
1391/2	0.19	0.17	–	–	–	0.46	0.29	–	0.12	–
84066/2	0.27	0.24	0.09	0.08	0.08	0.53	0.36	–	0.16	–
121.022/1	0.20	0.18	–	–	–	–	0.38	–	0.12	–
121.553/3	0.17	0.15	–	–	–	–	0.18	–	0.08	–
121.558/2	0.21	0.17	0.07	0.06	0.06	0.54	0.32	–	0.15	0.14
121.559/3	0.17	0.15	–	–	–	–	–	–	0.10	–
121.560/3	0.21	0.17	–	–	–	–	–	–	0.08	0.10
121.564/2	0.26	0.22	0.09	0.05	0.05	0.58	0.20	–	0.10	–
121.589/3	0.21	0.18	0.07	0.06	0.05	0.60	–	–	0.13	–
121.589/4	0.21	0.18	–	–	–	0.60	–	–	0.13	–
121.589/5	0.30	0.26	0.16	0.05	0.08	0.67	0.48	–	0.19	–
121.611	0.21	0.17	–	–	–	0.54	0.32	–	0.11	–
121.612	0.22	0.20	–	–	–	–	0.32	–	0.13	–
136.022	0.16	0.15	0.05	0.05	–	–	0.18	–	0.11	–

Table 30. Pygidial measurements of *Orometopus elatifrons.*

Specimen	X	X1	Y	Z	W	W1
872/5	0.09	0.05	0.12	0.14	0.44	–
1298/1	0.07	0.04	0.08	0.11	0.33	0.06
71041/7	0.21	0.14	0.31	0.35	0.92	0.06
121.622/2	0.08	0.04	0.10	0.20	–	–

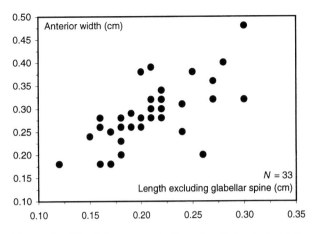

Fig. 73. Cranidia of *Orometopus elatifrons.* Length (sag.) of glabella excluding glabellar spine plotted against anterior width (tr.).

fers from the type species mainly in having the glabella expanding slightly anteriorly, lateral pits at the posterior part of the glabella, and an evenly curved, semielliptical anterior margin. *O. grypos* differs in having a strongly convex glabella, wider fixed cheeks, and an anteriorly diverging facial suture. A direct phylogenetical line between *O. elatifrons* and *O. primigenus* is conceivable, but the transition to *O. grypos* is not obvious.

The form described as *O. elatifrons* by Lake (1907) does not belong to the species. It differs in having a more parallel-sided glabella, wider (sag.) posterior parts of the fixed cheeks, smaller fixed cheeks opposite the palpebral lobes, longer anterior facial sutures, converging slightly anteriorly, merging with a more evenly curved, almost transverse anterior margin (Fig. 71B).

The reconstructions of the type species presented in the *Treatise on Invertebrate Paleontology* (Moore 1959) and in Fortey & Chatterton (1988, p. 212, Text-fig. 25A) are based on the English species, presumably Lake's (1907) specimens, and cannot be retained as representative for the type species.

Genus *Pagometopus* Henningsmoen, 1959

Type species. – Pagometopus gibbus Henningsmoen, 1959, pp. 170–171, Pl. 2:1–4, from the Bjørkåsholmen Formation at Bjørkåsholmen in Asker, Norway; by original designation of Henningsmoen (1959, p. 170).

Discussion. – The type species is so far the only representative of this genus. Henningsmoen (1959) pointed out the strong relationship between *Pagometopus* and *Orometopus*, but also noted similarities in the preglabellar field to that of the Family Hapalopleuridae Har-

quite unlike the preglabellar field of *O. elatifrons.* The anterior facial suture of *O. klouceki* and *O. grypos* are longer (sag.), converging anteriorly and marginally with a short, nearly transverse anterior border, thus having stronger affinity towards the facial suture of *Pagometopus* than to that of *O. elatifrons.*

Orometopus primigenus, O. elatifrons, and *O. grypos* are three closely related Scandinavian species with a progressively younger stratigraphical position. *O. primigenus* dif-

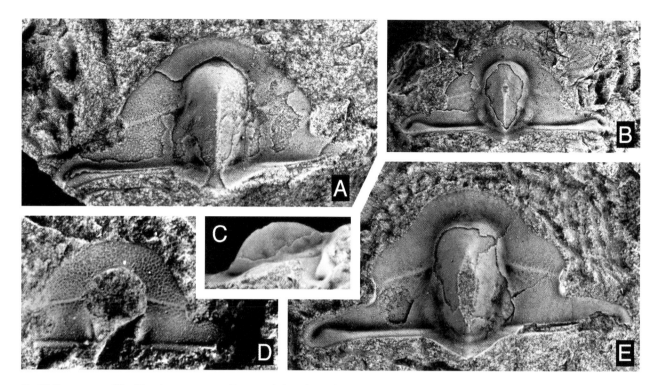

Fig. 74. Pagometopus gibbus Henningsmoen, 1959. □A. Dorsal view of cranidium, showing glabellar furrows; ×10, PMO 69579. Bjørkåsholmen, Asker. Coll.: G. Henningsmoen, 1958-11-23. Figured by Henningsmoen (1959, Pl. 2:3, 4). □B, C. Dorsal and lateral view of nearly complete cranidium; B ×6.5; C ×10, PMO 69578. Bjørkåsholmen, Asker. Coll.: G. Henningsmoen, 1959-11-23. Figured by Henningsmoen (1959, Pl. 2:2). □D. Dorsal view of internal mould of small cranidium showing pitted anterior part; ×12.5, PMO 1142/2. Found in a dark limestone nodule 28–32 cm above the base of the formation at Bjørkåsholmen, Asker. Coll.: Excursion 1910-09-24. □E. Dorsal view of holotype cranidium, showing anterior pits and eye ridges; ×9, PMO 69577. Bjørkåsholmen, Asker. Coll.: G. Henningsmoen, 1958-11-23. Figured by Henningsmoen (1959, Pl. 2:1).

rington & Leanza, 1957. Fortey & Shergold (1984) synonymized the Hapalopleuridae with the Orometopidae, but subsequently Fortey & Owens (1991) found that Hapalopleuridae should be united with Alsataspididae.

The genus *Skljarella* Petrunina, 1973, appears very similar to *Pagometopus* in many respects. The type species *Skljarella lidiae* Petrunina, 1973, Pl. 2:11, refigured by Fortey & Owens (1991, Fig. 10k), has the same general outline of the cranidium, but with a more rounded anterior margin and lateral fixed cheek pointing backwards laterally. The fixed cheeks of the British species *Skljarella cracens* Fortey & Owens, 1991, are similar to those of *Pagometopus*, but lack the deep lateral parts of the posterior border furrow. *Skljarella cracens* has an anterior margin similar to that of *Orometopus*, except for the anterior border brim.

The lateral position of the glabellar furrows are the same in both *Skljarella* and *Pagometopus*, and, as pointed out by Fortey & Owens (1991), the narrow preglabellar field is a uniting character within the family. A closer study of the group would be required for a comparison of generic characters and relationship.

Pagometopus gibbus Henningsmoen, 1959

Fig. 74

Synonymy. – □1959 *Pagometopus gibbus* n.sp. – Henningsmoen, p. 169, Pl. 2:1–4.

Holotype. – A cranidium (PMO 69577) from the Bjørkåsholmen Formation at Bjørkåsholmen in Asker, Norway. Identified and figured by Henningsmoen (1959, p. 169, Pl. 2:1).

Material. – Seven more-or-less incomplete cranidia, the largest (PMO 69577, Fig. 74E) 0.53 cm long (sag.).

Discussion. – A diagnosis and description were presented by Henningsmoen (1959), and although new material has been recognized, a redescription is not necessary here.

The close resemblance to species of *Orometopus* is evident in the shape and outline of the glabella, the posterior part of the fixed cheeks, and in the palpebral lobes. The main difference is in the shape of the anterior facial suture of *Pagometopus*, which converges strongly anteri-

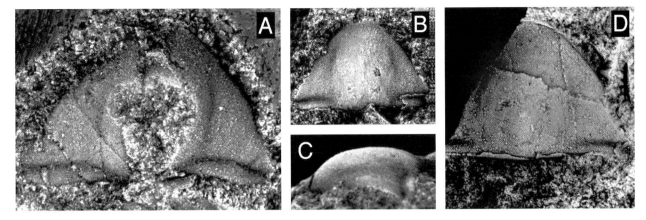

Fig. 75. Falanaspis aliena Tjernvik, 1956. □A. Dorsal view of large, incomplete cephalon; ×12, PMO 121.656. Found in the upper part of the formation at Øvre Øren, Modum. Coll.: J.O.R. Ebbestad, 1991-07-22. □B, C. Dorsal and lateral view of cephalon; ×12, PMO 136.016. Found 10–18 cm above base of the formation at Færdenveien in Klekken, Ringerike. Coll.: J.O.R. Ebbestad, 1992-09-06. □D. Dorsal view of cephalon showing lateral margin; ×12, PMO S1238. Vækerø, Oslo. Coll.: L. Størmer, 1919. Figured by Henningsmoen (1959, Pl. 1:8).

orly and merges with a narrow (tr.), rounded anterior margin (Fig. 74E).

A small specimen attributed to this species (Fig. 74D) was found as an internal mould in a dark limestone nodule from near the base of the formation. It appears slightly different, with prominent reticulated pattern in front of the eye ridges and with a more rounded outline of the facial suture. The size and ontogenetic variations would account for these differences.

Orometopus klouceki Vaněk, 1965, has a comparable anterior facial suture, converging anteriorly and merging with a short (tr.) anterior border. However, it does not have the distinct tapering outline and flat preglabellar field seen in *Pagometopus*.

Family Alsataspididae Turner, 1940

Genus *Falanaspis* Tjernvik, 1956

Type species. – *Falanaspis aliena* Tjernvik, 1956, pp. 272–274, Pl. 11:19–21, from the Biozone of *Megistaspis armata* (lower Arenig) at Stenbrottet in Västergötland, Sweden; by original designation of Tjernvik (1956a, p. 272).

Discussion. – Two species are known; the type species from the lower and lowermost Middle Ordovician biozones of *Apatokephalus serratus*, *Megistaspis armata* and *M. planilimbata* in Norway and Sweden, and *Falanaspis extensa* Fortey, 1975, from the Middle Ordovician Olenidsletta Member of Svalbard. *Falanaspis* does bear some resemblance to the genus *Seleneceme* Clark, 1924. An emended diagnosis of the family was given by Fortey & Shergold (1984).

Falanaspis aliena Tjernvik, 1956

Fig. 75

Synonymy. – □1956a *Falanaspis aliena* n.sp. – Tjernvik, pp. 272–274, Text-fig. 44, Pl. 11:19–21. □1959 *Falanaspis aliena* Tjernvik – Harrington *et al.* in Moore, p. O428. □1959 *Falanaspis aliena* Tjernvik – Henningsmoen, p. 171, Pl. 1:8. □1984 *Falanaspis aliena* Tjernvik – Fortey & Shergold, p. 352.

Holotype. – A cranidium (PMU Vg 389) from the Biozone of *Megistaspis armata* (lower Arenig) at Stenbrottet in Västergötland, Sweden. Identified and figured by Tjernvik (1956a, Pl. 11:20).

Norwegian material. – Four cranidia, the largest 0.33 cm long (sag.). Table 31 shows the measurements of the available material.

Discussion. – The species was described by Tjernvik (1956a), and a redescription is not necessary here, except that the outline of the glabella appears somewhat more elongated (sag.) in the Norwegian specimens.

The Swedish specimens of *F. aliena* occur in the Biozone of *Megistaspis armata*, overlying the Biozone of *Apatokephalus serratus*. Norwegian specimens of *F. aliena* (PMO 121.656, Fig. 75A, and PMO 136.087)

Table 31. Cephalic measurements of *Falanaspis aliena*.

Specimen	A	B	J	K1
S1238	0.26	0.23	0.48	0.16
121.656	0.31	0.28	0.50	0.22
136.016	0.23	0.21	0.33	0.15

were found in the upper part of the Bjørkåsholmen Formation at Modum (see earlier fauna log and discussion of this section) and in the lower part of the overlying Tøyen Formation at Vestfossen. At both localities the species was associated with a trilobite fauna of the Biozone of *Megistaspis planilimbata* (Hoel, in press).

The remaining two specimens (Fig. 75B–D) are associated with a trilobite fauna of the Biozone of *A. serratus* and are thus older than the type material, which is only referred to the Biozone of *Megistaspis armata*. They are, however, virtually indistinguishable, and there seems to be no reason to describe them as separate forms. This means that the known stratigraphical range of this species is extended both downwards and upwards.

Family Leiostegiidae Bradley, 1925

Genus *Agerina* Tjernvik, 1956

Type species. – *Agerina erratica* Tjernvik, 1956, pp. 197–199, Text-fig. 29, Pl. 1:24–26, from the upper Planilimbata Limestone (Arenig) at Örebro in Närke, Sweden; by original designation of Tjernvik (1956a, p. 197).

Discussion. – Fortey & Shergold (1984, p. 322) included *Agerina* in the family Leiostegiidae because the glabella is like the glabella of *Annamitella*, also assigned to the same family. Ingham & Tripp (1991) followed their view, but with reservations.

Ludvigsen (1980) discussed and listed the species assigned to this genus. To the list can be added *A. laurentica* Ingham & Tripp, 1991, from Girvan, Scotland (Llandeilo). The genus ranges from early Tremadoc to earliest Caradoc and is found in Argentina, Canada, China, Scandinavia, Scotland, Russia and Turkey.

Brackebuschia Harrington & Leanza, 1957, was regarded a junior synonym of *Agerina* by Ludvigsen (1980, p. 99), based on the great similarity with species of *Agerina*. He also suggested that low palaeolatitude *Agerina* species were restricted to cold water sites on the slope and platform-edge.

Agerina praematura Tjernvik, 1956

Fig. 76

Synonymy. – □1906 *Orometopus elatifrons* (Angelin) – Moberg & Segerberg, p. 99, Pl. 7:4, 5. □1956a *Agerina praematura* n.sp. – Tjernvik, p. 200, Pl. 1:22, 23. □1973 *Agerina praematura* Tjernvik – Dean, p. 305. □1975 *Agerina praematura* Tjernvik – Lu, p. 185. (p. 391, English version). □1980 *Agerina praematura* Tjernvik – Ludvigsen, pp. 99–100.

Holotype. – A cranidium (PMU Vg 267) from the Bjørkåsholmen Formation at Stenbrottet in Västergötland, Sweden. Identified and figured by Tjernvik (1956a, Pl. 1:22).

Norwegian material. – One cranidium, 0.26 cm long (sag.) and one pygidium, 0.18 cm long (sag.).

Remarks. – Tjernvik (1956a, p. 200) provided a diagnosis and a short description of this species. A few modifications are needed, however, and an emended diagnosis and description are given here.

Emended diagnosis. – Cephalon with genal spines. Glabella rectangular, outlined by very prominent occipital, dorsal and preglabellar furrows. The latter cross the anterior facial sutures opposite the anterolateral corners of the glabella. All glabellar furrows positioned within area covered by sagittal length of palpebral lobes. Pygidial rachis with three rachial rings. Terrace ridges on pygidial border only.

Emended description. – Sagittal length of cranidium slightly less than posterior width. Glabella subrectangular, wide (tr.), making up half the posterior width, convexity (tr.) low, curving almost vertically laterally (Fig. 76D), slanting anteriorly. Occipital ring tapering laterally, outlined anteriorly by deep, transverse occipital furrow. The 1S furrows are situated marginally just anteriorly of posterior extremity of eyes, curving sharply backwards. The 2S furrows are situated marginally, slightly posterior of anterior extremities of eyes, transverse. The 3S furrows are situated opposite to anterior extremities of eyes, converging slightly obliquely forward. Anterior glabellar lobe well rounded in front. Dorsal furrow distinct. Posterior part of fixed cheeks short (tr. and sag.) with distinct posterior border furrow. Palpebral lobes semicircular, situated close to and slightly posterior to middle of glabella, making up one-third of sagittal length. Anterior facial suture and anterior margin subparallel to anterior glabellar lobe. Preglabellar furrow deep, continued laterally, outlining wide (sag.) anterior border with terrace ridges.

Anterior width of pygidium slightly less than twice sagittal length. Rachis convex (tr.), tapering backwards, transverse posteriorly, with three distinct rachial rings. Pleural fields crossed by two pairs of transverse interpleural furrows and pleural furrows, all faint except anterior pleural furrow. Faint border furrow present marginally. Posterior margin semicircular with terrace ridges, anterior margin transverse adrachially, articulating facets directed obliquely backwards laterally (Fig. 76B, C).

Discussion. – The glabellar furrows are barely visible in the Norwegian material, but the great similarities in size and morphology compared with the Swedish material give no reason to view this as a different form.

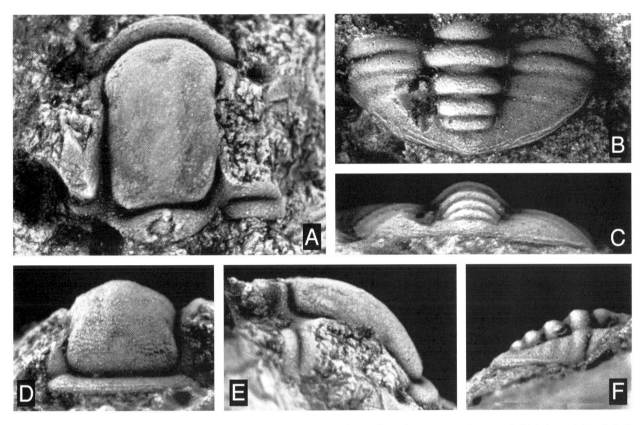

Fig. 76. Agerina praematura Tjernvik, 1956. □A, D, E. Dorsal, anterior and lateral view of cranidium; ×20, PMO 70721. Bjørkåsholmen, Asker. Coll.: F. Nikolaisen, 1969-05-13. □B, C, F. Dorsal, posterior and lateral view of pygidium; ×20, PMO H2625/2. Vestfossen, Øvre Eiker. Coll.: W.C. Brøgger, 1879.

The glabella of *A. praematura* closely resembles that of *A. pamphylica* Dean, 1973, both in proportions and glabellar segmentation, and that of *A. elongata* Lu, 1975, in the proportions of the glabella and position of the palpebral lobes. The pygidium of *A. praematura* resembles that of *A. erratica* Tjernvik, 1956, and *A. pamphylica*, but the latter has four rachial rings, and the posterior margin of the type species is slightly pointed sagittally.

Family Harpididae Raw, 1949

Genus *Harpides* Beyrich, 1846

Type species. – Harpides hospes Beyrich, 1846, Pl. 4:4, Tremadoc strata in Bohemia, Czech Republic, by original designation of Beyrich (1846).

Discussion. – By the ruling of the International Commission on Zoological Nomenclature (1987, opinion 1436), the subfamily Harpidinae Raw, 1949, based on *Harpides* Beyrich, 1846, was placed on the Official List of Family-Group Name in Zoology together with Harpetidae Hawle & Corda, 1847. The latter was formerly known as Harpi-

dae but was considered a homonym of the molluscan family by the same name.

The species belonging to this genus were listed by Lisogor (1961), Poulsen (1965), Wolfart (1970) and Peng (1990).

Pillet & Courtessole (1980) erected four subgenera of *Harpides*: *H. (Harpides)*, based on *H. hospes* Beyrich, 1846; *H. (Dictyocephalites)*, based on *H. villebruni* (Bergeron, 1895); *H. (Paraharpides)*, based on *H. atlanticus* Billings, 1865; *H. (Metaharpides)* based on *H. neogaeus* Harrington & Leanza, 1957.

The subdivision of *Harpides* must, however, be treated with some caution.

H. rugosus Sars & Boeck, 1838, was assigned to *H. (Paraharpides)* by Pillet & Courtessole (1980). It shows a highly convex cephalon with a well-defined cephalic border (Fig. 69A). These characters separate it from the assigned subgenus, but do not readily place it within the remaining three. The cephalon is semicircular without transverse genal spines or large backward-curving spines, and cannot be assigned to either *H. (Dictyocephalites)* or *H. (Metaharpides)*. It has very prominent alae, and the border is not obviously concave, thus excluding it from *H. (Harpides)*.

H. grimmi Barrande, 1852, assigned to *H.* (*Dictyocephalites*), is so similar to *H. rugosus* that it has been suggested to be conspecific (Raw 1949; Henningsmoen 1959). The diverging genal spines, somewhat suspect in the drawing of Pillet & Courtessole (1980, p. 415, Fig. 1:3), and the rather angular outline of the cephalon distinguish it from *Paraharpides* and *H. rugosus* (Pillet & Courtessole 1980).

The convex cephalon of *H. neogaeus*, assigned to *H.* (*Metaharpides*), has a wide (tr.) border, carrying distinct radiating ridges and is remarkably similar to the cephalon of the type species, *H. hospes*, assigned to *H.* (*Harpides*). The distinctive feature is the long, backward-pointed, curved genal spines seen in *H. neogaeus*.

Raw (1949, p. 511), Whittington (1950, p. 302) and Henningsmoen (1959, p. 169) suspected that the genal spine seen in *Harpides* extends from the doublure, the cephalon having a marginal suture. This feature is seen both in *H. grimmi* and *H. rugosus* (Fig. 68H), and may relate to the whole genus. Thus, the morphology of the genal spines may be independent of the rather conservative morphology of the cephala of the *Harpides*.

Consequently, the subgeneric division of *Harpides sensu* Pillet & Courtessole (1980) is here regarded invalid. It fails to consider the importance of the doublure and is based on too few characters, leaving out the conspicuous cephalic structures, the position of the occipital tubercle, the palpebral tubercles, and the glabellar structures.

Harpides rugosus (Sars & Boeck, 1838)

Figs. 77, 78

Synonymy. – □1838 *Trilobites rugosus* Sars & Boeck mscr. – Boeck, p. 143. □1854 *Harpides rugosus* (Boeck) – Angelin, p. 87, Pl. 41:7, 7a. □1869 *Harpides rugosus* (Boeck) – Linnarsson, p. 67. □1882 *Harpides rugosus* (Boeck) – Brøgger, p. 127. □1906 *Harpides rugosus* (Boeck) – Moberg & Segerberg, p. 85; Pl. 5:3–5. □1906 *Harpides rugosus* (Boeck) – von Post, Figs. 3–5. □1940 *Harpides rugosus* (Boeck) – Størmer, p. 146, Pl. 1:14, 15. □1949 *Harpides rugosus* (Boeck) – Raw, p. 511, Figs. 1, 2. □1950 *Harpides rugosus* (Boeck) – Whittington, p. 302. □1956a *Harpides rugosus* (Boeck) – Tjernvik, p. 268. □1959 *Harpides rugosus* (Boeck) – Harrington *et al.* in Moore, p. O418, Fig. 321:3a. □1959 *Harpides rugosus* (Boeck) – Henningsmoen, p. 166, Pl. 2:5–11. □?1961 *Harpides rugosus conicus* (Boeck) – Lisogor, p. 81, Pl. 3:6, 7. □1965 *Harpides rugosus* (Boeck) – Whittington, p. 311. □1965 *Harpides rugosus* (Boeck) – Poulsen, p. 96. □1980 *Harpides* (*Paraharpides*) *rugosus* (Boeck) – Pillet & Courtessole, pp. 415–416, Fig. 1:6, Fig. 2a. □1984 *Harpides rugosus* (Boeck) – Mergl, p. 30. □1989 *Harpides rugosus* (Boeck) – Dean, p. 16. □1990 *Harpides rugosus* (Boeck) – Peng, p. 111.

Lectotype. – A fragmentary cranidium (PMO 20053) from the Bjørkåsholmen Formation in Oslo, Norway. Selected and figured by Størmer (1940, p. 146, Pl. 1:14).

Norwegian material. – Fourteen more-or-less incomplete cranidia and two fragmentary doublure plates.

Remarks. – The species has been properly described by Henningsmoen (1959), but has not previously been diagnosed.

Diagnosis. – Cephalon semicircular, convex, glabella tapering forward, with three pairs of lateral glabellar furrows, occipital furrow curving forward in front of median occipital node. Eye tubercles situated slightly in front of glabella. Preglabellar boss distinct but not well defined, radiating ridges thicker marginally, without fine meshwork of ridges between.

Discussion. – The description given by Henningsmoen (1959) is followed here, except that the eye tubercles are positioned just forward of a transverse line through the front of glabella. New material shows an unusual high convexity of the cephalon, being horizontal near the glabella, and curving steeply down anteriorly and laterally with a slightly concave transition to the border (Fig. 77B, C). The transition is also reflected by the change in the radiating ridges, becoming thicker marginally and losing much of the fine anastomosing ridges between them (Fig. 78A). The occipital furrow of young specimens is effaced medially, the median occipital tubercle is distinct, the eye ridges and eyes are almost effaced, and small tubercles dominate the surface structure (Fig. 78B, C).

Henningsmoen (1959) was the first to illustrate the nature of the doublure in *Harpides*. Whittington (1965, pp. 309–312, Pl. 6:2–4) discussed the doublure found in *H. atlanticus* from Newfoundland and demonstrated the similar structures of girders crossed by caeca and lines of pits found in *H. rugosus*. The cephalon of *H. atlanticus* is quite similar to that of *H. rugosus*, but differs in having a tubercle in front of the occipital ring, and eye ridges with large tubercles positioned behind a transverse line through the front of the glabella. The thick radiating ridges are positioned closer to the glabella and are effaced laterally.

H. grimmi Barrande, 1852, from Bohemia is strikingly similar to the Norwegian species, and Raw (1949) and Henningsmoen (1959) suggested that they might be conspecific. Mergl (1984, p. 30) and Peng (1990, p. 111), however, showed that the two species were different, and it is also noted here that *H. rugosus* has a distinct forward curve in front of the median occipital tubercle, the L1 lobes are not isolated by the 1S furrows, and the doublure has radiating ridges and tubercles.

H. nodorugosus Poulsen, 1965, is another closely related species, but differs slightly in having the palpebral tuber-

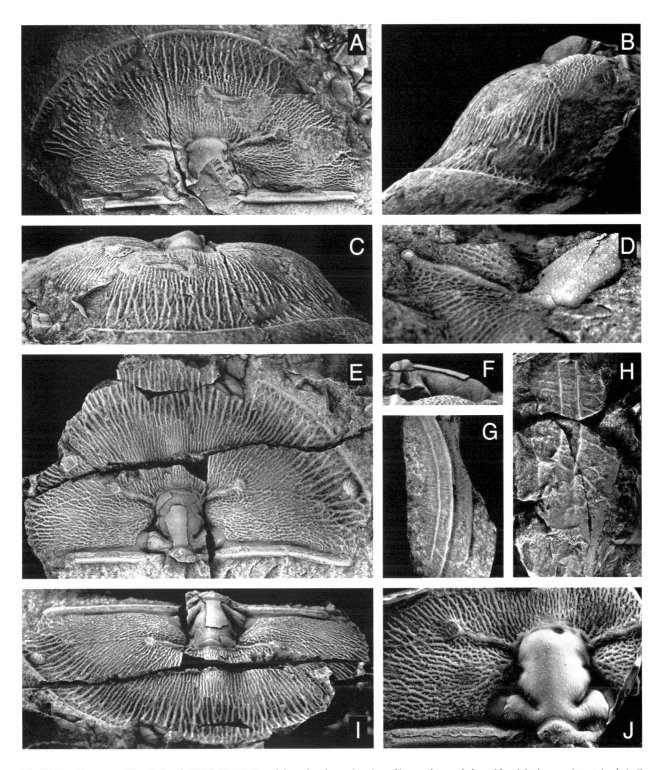

Fig. 77. Harpides rugosus (Sars & Boeck, 1838). □A–C. Dorsal, lateral and anterior view of incomplete cephalon with original convexity retained; A, C ×1.5; B ×2, PMO 70510/1. Bjørkåsholmen, Asker. Coll.: D.L. Bruton, 1968-09-10. □D. Detail of eye and eye ridge; ×5.5, PMO 6175/1. Vestfossen, Øvre Eiker. Coll.: W.C. Brøgger, 1879. □E, F, I. Dorsal and frontal view of cephalon and detail of glabella, lateral view; E ×1.75; F ×2.5, PMO 69583. Bjørkåsholmen, Asker. Coll.: G. Henningsmoen, 1958. Figured by Henningsmoen (1959, Pl. 2:11). □G. Latex replica from external mould of cephalic doublure, ventral view; ×1.5, PMO 1290. Bjørkåsholmen, Asker. Coll. unknown, 1915. Figured by Henningsmoen (1959, Pl. 2:6). □H. Ventral view of cephalic doublure; ×4, PMO 56024/1. Bjørkåsholmen, Asker. Coll.: L. Størmer, 1934-05-03. Figured by Henningsmoen (1959, Pl. 2:10). □J. Latex replica from external mould of glabellar area; ×3, PMO 69584. Bjørkåsholmen, Asker. Coll.: G. Henningsmoen, 1958.

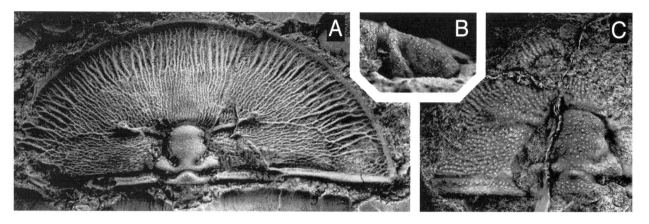

Fig. 78. Harpides rugosus (Sars & Boeck, 1838). □A. Dorsal view of complete cephalon; ×1.75, PMO 83825. Bjørkåsholmen, Asker. Coll.: F. Nikolaisen, 1968–07-12. □B, C. Lateral and dorsal view of latex cast of small cranidium; ×4, PMO 35948. Steinsodden, Ringsaker. Coll.: T. Strand, 1926.

cles positioned behind the front of the glabella, no preglabellar boss, and a preoccipital tubercle.

The Chinese species *H. troedssoni* Chang & Fan, 1960, is very similar to *H. rugosus*, having the same outline of the cephalon and similar definition of the border and the cephalic structure. The palpebral tubercles are more similar to those of *H. grimmi*.

Family Pilekiidae Sdzuy, 1955

Subfamily Pilekiinae Sdzuy, 1955

Genus *Parapilekia* Kobayashi, 1934

Type species. – Calymene? speciosa Dalman, 1827, p. 100 [285], from the Tremadoc series on Öland, Sweden; subsequently designated by Holliday (1942, p. 475).

Discussion. – Jell & Stait (1985) revived the family status of Pilekiidae. Their views were discussed and followed by Peng (1990) and are also adopted here. Species belonging to this genus were listed by Fortey (1980) and Peng (1990). To this can be added *Parapilekia ferrigena* Mergl, 1994, from the Tremadoc of Central Bohemia, Czech Republic.

Sdzuy (1955) regarded *Parapilekia* a junior subjective synonym of *Pilekia* Barton, 1915, and was supported by Vaněk (1965), Wolfart (1970) and Lane (1971). Fortey (1980), however, distinguished *Pilekia* by the strongly tapering glabella and the tumid glabellar lobes. The latter view is followed here.

Peng (1990) erected a new genus *Sinoparapilekia* in a new subfamily Sinoparapilekiinae. This genus contains *Parapilekia*-like species, mainly from China, distinguished mainly by having a distinct posterior area of the fixed cheeks. *P. jacquelinae* Fortey, 1980, was assigned to this new subfamily by Peng (1990, p. 113).

The holaspid stages of *P. afghanensis* (Wolfart, 1970) have a small anterior area of the fixed cheeks. Other characteristics of its morphology correspond well with the diagnosis of *Sinoparapilekia* Peng, 1990, and suggest that it belongs in this genus.

Parapilekia speciosa (Dalman, 1827)

Figs. 79, 80

Synonymy. – □1827 *Calymene? speciosa* n.sp. – Dalman, p. 75 [260]. □1835 *Calymene speciosa* Dalman – Sars, p. 339. □1854 *Cyrtometopus speciosus* (Dalman) – Angelin, p. 77, Pl. 39:7, 7b. □1882 *Cheirurus foveolatus?* Angelin – Brøgger, p. 130, Pl. 2:5. □1888 *Cyrtometopus speciosus* (Dalman) – Lindström, p. 11. □1906 *Cyrtometopus speciosus* (Dalman) – Moberg & Segerberg, p. 103, Pl. 7:15–17. □1934 *Parapilekia speciosa* (Dalman) – Kobayashi, pp. 569–570. □1942 *Parapilekia speciosa* (Dalman) – Holliday, p. 475. □1959 *Parapilekia speciosa* (Dalman) – Harrington *et al. in* Moore, p. O441, Fig. 346:8a, b. □?1961 *Protopliomerops speciosus* (Dalman) – Balashova, p. 138, Pl. 3:10. □1970 *Pilekia speciosa* (Dalman) – Wolfart, p. 62. □1980 *Parapilekia speciosa* (Dalman) – Fortey, pp. 80–81.

Holotype. – A cranidium (RM Ar23163) from the Bjørkåsholmen Formation at Öland, Sweden. Described by Dalman (1827, p. 75 [260]). By monotypy. Figured by Angelin (1854, Pl. 39:7a, b).

Norwegian material. – One almost complete specimen, three fragmentary cranidia and one pygidium. Tables 32 and 33 present measurements of the cranidia and pygidium, respectively.

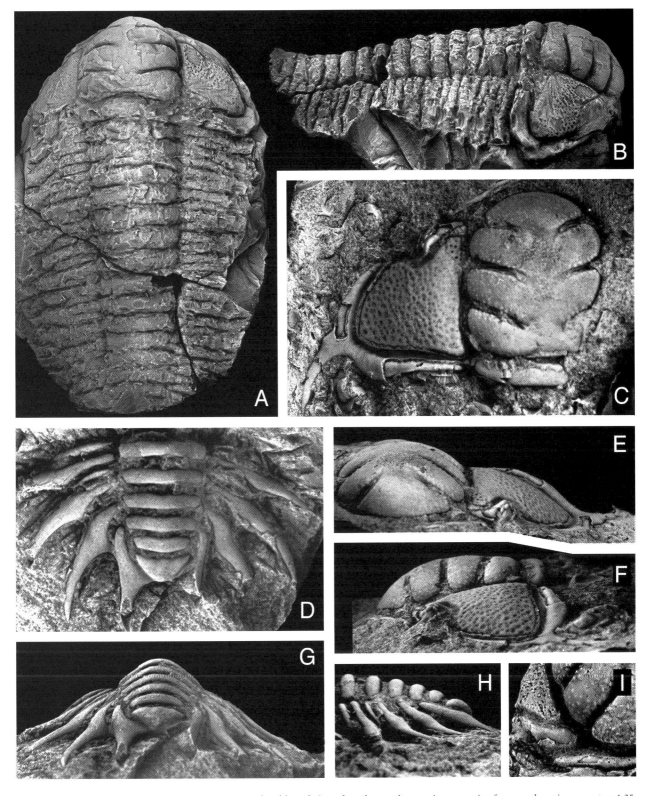

Fig. 79. Parapilekia speciosa (Dalman, 1827). □A, B. Dorsal and lateral view of nearly complete specimen carrying fourteen thoracic segments; ×1.25, PMO 136.070. Vang skole at Klekken, Ringerike. Coll.: A. Saga, 1972. □C, E, F. Dorsal, anterior and lateral view of latex replica from external mould of cranidium; ×2.5, PMO 1302/5. Bjørkåsholmen, Asker. Coll. unknown, 1915. □D, G, H. Dorsal, posterior and lateral view of pygidium, showing the prominent pygidial spines; ×2.5, PMO 20111. Ramtonholmen, Røyken. Coll.: Stud. Samuelsen. Figured by Brøgger (1882, Pl. 2:5). □I. Dorsal view of incomplete cranidium, showing anterior margin; ×2.5, PMO 84111. Bjørkåsholmen, Asker. Coll.: L. Størmer, 1955-04-29.

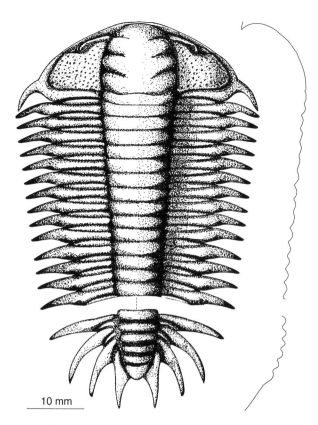

Fig. 80. Reconstruction of *Parapilekia speciosa* with fourteen thoracic segments.

Remarks. – *Parapilekia speciosa* has not been revised since Moberg & Segerberg (1906). A diagnosis and emended description are presented here.

Diagnosis. – Posterior width of cranidium (excluding genal spines) twice sagittal length. Palpebral lobes opposite 3P glabellar lobes, close to glabella (tr.) on flat part of fixed cheeks. Posterior fixed cheeks broad (sag.). Posterior margin hook-like laterally, curving into genal spines. Pygidium with parallel-sided rachis. Pygidial spines distinctly spaced, robust.

Emended description. – Posterior width of cranidium twice sagittal length (excluding genal spines). Glabella large, convex (tr.), parallel-sided, expanding slightly forwards, occupying nearly full sagittal length and two-fifths of posterior width, rounded in front. Occipital ring narrow (sag.), outlined by distinct occipital furrow curving slightly crescentically sagittally, obliquely forwards laterally. Three pairs of deep glabellar furrows, directed slightly obliquely backwards from the dorsal furrow at 15°, all reaching equally deep adrachially, separated by equally large (sag.) glabellar lobes. Anterior glabellar lobe semicircular in outline, narrow (tr.). Dorsal furrow distinct. Prepalpebral area of fixed cheeks large, subquadratic in outline, curved steeply down laterally, extending backwards from opposite 2S furrows.

Posterior margin directed slightly obliquely backwards near glabella, curving markedly forward laterally to genal angle. Posterior border furrow directed slightly obliquely forwards, curving forwards and slightly inwards at genal angle (Fig. 79C). Short, curved genal spines extend backwards from genal angle at different angle to semicircular cephalic margin. Posterior facial suture extending nearly transverse from palpebral lobes, cutting margin opposite 2S furrows. Palpebral lobes small, situated close to glabella, opposite 3L glabellar lobes, raised above fixed cheeks, outlined by distinct palpebral lobe furrows. Eye ridges short, directed obliquely forwards, reaching dorsal furrow opposite of 3S furrows (Fig. 79I). Preglabellar furrow distinct, crossing anterior facial suture directly in front of eye ridges, outlining raised anterior border. Anterior facial suture converging anteriorly to meet short (tr.) anterior margin. Glabella covered with small tubercles, fixed cheeks having a reticulated structure of small, irregularly spaced grooves.

Thorax with at least fourteen segments (Fig. 79A, B). Rachis tapering backwards, pleural fields wider posteriorly. Rachial ring convex (tr.), parallel-sided, pleurae slightly convex (sag.), pointed laterally. Pleural furrows transverse adrachially, curving backwards, cutting posterior margin of pleurae at slightly more than half the length (tr.).

Pygidium wider (tr.) than long (sag.). Rachis convex (tr.), parallel-sided with five distinct rachial rings. Rachial furrows wide (sag.). Rachial termination subtriangular, extending vertically down to margin (Fig. 79H). Pleural fields narrow (tr.) with four pairs of convex (sag.) pleurae. Deep pleural furrows present on anterior pleurae, progressively less distinct and more curved posteriorly, absent on posterior pleurae. Pleurae separated by distinct interpleural furrows. Pleurae continued into pointed spines laterally, directed nearly transversely near anterior margin, directed backwards posteriorly. Deep furrow posterior of nearly transverse anterior margin.

Discussion. – Specimens of this species are extremely rare. The few new Norwegian specimens demonstrate the high convexity of the cephalon and, for the first time, the nature of the thoracic region.

Table 32. Cranidial measurements of *Parapilekia speciosa.*

Specimen	A	B	C	C1	C2	J	J1	J2	K	K1
1302/5	2.03	1.95	0.46	0.31	1.18	4.60	0.90	1.22	1.60	1.03
84111	–	–	0.62	–	–	–	2.88	–	–	1.50
136.070	–	2.16	–	0.39	1.46	4.37	–	–	2.04	1.38

Table 33. Pygidial measurements of *Parapilekia speciosa.*

Specimen	X	X1	Y	Z	W
20111	0.84	0.55	1.28	1.28	1.77

The cranidium and pygidium of *P. olesnaensis* (Růžička, 1934) are very similar to those of *P. speciosa*. The glabella is proportionally shorter, and the posterior part of the fixed cheeks is shorter (sag.). The pygidium differs in having a shorter (sag.), more tapering rachis, and carrying more bristly pygidial spines. The cranidium of *P. latilus* (Liu *in* Zhou *et al.*, 1977) has a proportionally narrower (tr.) glabella, palpebral lobes positioned further laterally, and a more prominent anterior border than that of *P. speciosa*. It is interesting that this species also has at least thirteen thoracic segments (see Peng 1990, Pl. 21:6a). Species like *P. sougyi* Destombes *in* Destombes *et al.*, 1969, and *P. atecea* Hammann, 1974, differ somewhat from the type species in having a more rounded glabella, not entirely unlike that of *Pilekia*. A reconstruction of *Parapilekia speciosa* is shown in Fig. 80.

Subfamily Sinoparapilekiinae Peng, 1990

Genus *Pliomeroides* Harrington & Leanza, 1957

Type species. – Protopliomerops deferrariisi Harrington, 1938, p. 184, Text-fig. 6, Pl. 6:13, 19, 21, 23, from the *Notopeltis orthometopaz* Biozone (upper Tremadoc) at the west side of Quebrada de Humahuaca in the Tumbaya district, Jujuy province, Northern Argentina; subsequently designated by Harrington & Leanza (1957, p. 218).

Discussion. – The genus was originally placed in the family Pliomeridae Raymond, 1913. However, *Pliomeroides* differs from all Pliomeridae in having a distinct anterior area of the fixed cheeks, and Peng (1990, p. 113, 114) assigned it to the Subfamily Sinoparapilekiinae Peng, 1990, of the Family Pilekiidae Sdzuy, 1955. The family was discussed and revived by Jell & Stait (1985).

Balashova (1966) proposed *Evropeites* as a subgenus to *Pliomeroides*. The poorly known *Pliomeroides primigenus* var. *lamanskii* was designated as type species. Fortey (1980) found that the cephalic and pygidial characters of the subgenus justified a generic status for *Evropeites*, including all material formerly assigned to *Pliomeroides* except the type species. Based on new photographs of the *Evropeites* type species, Fortey & Droser (1996, p. 97) transferred the earlier described *E. hyperboreus* from Svalbard (Fortey 1980) to the genus *Pseudomera* Holliday, 1942. *Pliomeroides* is similar to *Pseudomera* in many respects, but the latter genus lacks the anterior area of fixed cheeks distinguishing the two families, as discussed above.

Here *Pliomeroides* is taken to include *P. primigenus* (Angelin, 1854), in addition to the type species, while *Evropeites* would include its type species. *Pliomeroides* is

distinguished from *Evropeites* by the tapering glabella, with the 3S glabellar furrows reaching the dorsal furrow, palpebral lobes situated opposite 2P lobes, and a pygidium with distinct pleural and interpleural furrows. Illustrations of the type species of *Pliomeroides* in Moore (1959, p. O443, Fig. 347:9a) are not ideal. However, it is clear from the original descriptions and pictures that the pygidium resembles those of *Evropeites* more closely than suggested by Fortey (1980, p. 89). It is not clear whether the pygidial rachis of the type species has a pair of pits on the rachial termination or not.

Pliomeroides primigenus (Angelin, 1854)

Figs. 81–83

Synonymy. – ☐1854 *Pliomera primigena* n.sp. – Angelin, p. 90, Pl. 41:15. ☐1869 *Pliomera primigena* Angelin – Linnarsson, p. 62, Pl. 1:10. ☐1882 *Amphion primigenus* (Angelin) – Brøgger, p. 134. ☐1906 *Cyrtometopus primigenus* (Angelin) – Moberg & Segerberg, p. 101, Pl. 7:13, 14, *non* Fig. 12. ☐1934 *Protopliomerops primigenus* (Angelin) – Kobayashi, pp. 570–571. ☐1937 *Protopliomerops primigenus* (Angelin) – Harrington, p. 120, Pl. 5:2, 3. ☐1957 *Pliomeroides primigenus* (Angelin) – Harrington & Leanza, p. 219. ☐1966 *Pliomeroides* (*Evropeites*) *primigenus* (Angelin) – Balashova pp. 18–19. ☐1980 *Evropeites primigenus primigenus* (Angelin) – Fortey, pp. 88–89.

Type material. – A pygidium from the Bjørkåsholmen Formation in Oslo, Norway. Described and figured by Angelin (1854, Pl. 41:15). The type specimen was not found in the type collections at the Swedish Museum of Natural History, Stockholm, and seems to be lost. Until a neotype is selected or the type material recovered, the Norwegian material must suffice to define the taxon.

Material. – One nearly complete specimen, nineteen cranidia, one free cheek and eleven pygidia. The largest cranidium is 1.54 cm long (sag.), while the largest pygidium is 0.77 cm long (sag.). Tables 34 and 35 present measurements of the cranidia and pygidia, respectively.

Remarks. – This species has not been redescribed since Moberg & Segerberg (1906), and is poorly understood. A diagnosis and full description is provided here.

Diagnosis. – Sagittal length of cranidium one-third of posterior width. Glabella with three pairs of deep furrows, tapering slightly anteriorly. Transition to fixed cheeks defined by steep slope and thin ridge along glabella. Genal angle slightly pointed. Anterior border furrow deep sagittally and laterally. Deep pits at anterolateral corners. Pygidium with slender, tapering rachis with five rachial rings and paired pits at terminal rachial lobe. Five pairs of spines directed obliquely downwards and rearwards.

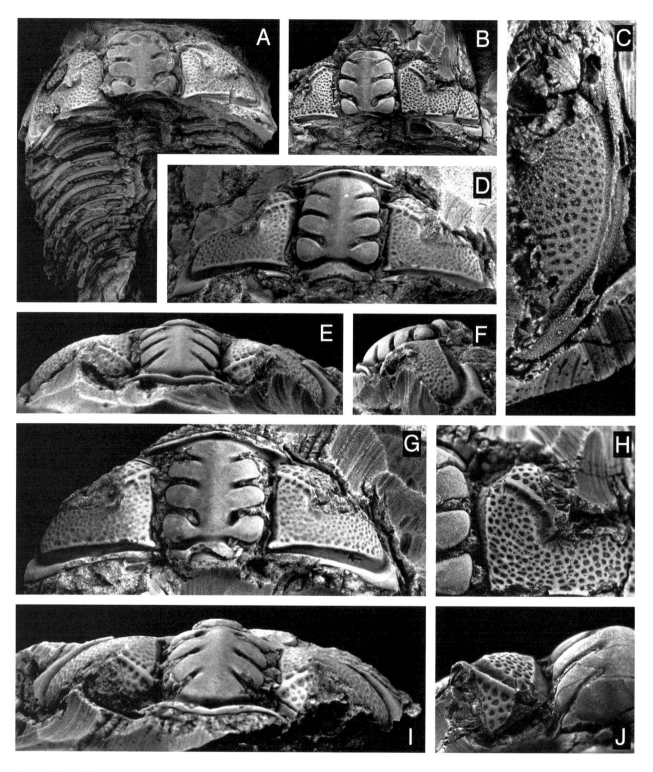

Fig. 81. Pliomeroides primigenus (Angelin, 1854). □A. Dorsal view of nearly complete specimen carrying ten thoracic segments; ×1.75, PMO 121.539/1. Bjørkåsholmen, Asker. Coll.: Excursion 1960. □B. Dorsal view of cranidium, showing the prominent structures on the fixed cheeks (internal mould); ×1.5, PMO 84030/1. Bjørkåsholmen, Asker. Coll.: G. Henningsmoen, 1958-11-16. □C. Dorsal view of right free cheek, showing facial suture; ×4, PMO 33167. Nordre Grefsen at Gran, Hadeland. Coll.: T. Münster, 1893-08-31. □D, E, F. Dorsal, anterior and lateral view of cranidium, showing anterior pits and anterior border; ×3, PMO 1394. Bjørkåsholmen, Asker. Coll. unknown, 1915. □G, I. Dorsal and anterior view of well-preserved cranidium; ×2.5, PMO 83828/1. Bjørkåsholmen, Asker. Coll.: F. Nikolaisen, 1968-07-12. □H, J. Dorsal and anterior view of fixed cheek, showing eye ridge and nearly complete palpebral lobe; ×4, PMO 84030/1. Bjørkåsholmen, Asker. Coll.: G. Henningsmoen, 1958-11-16.

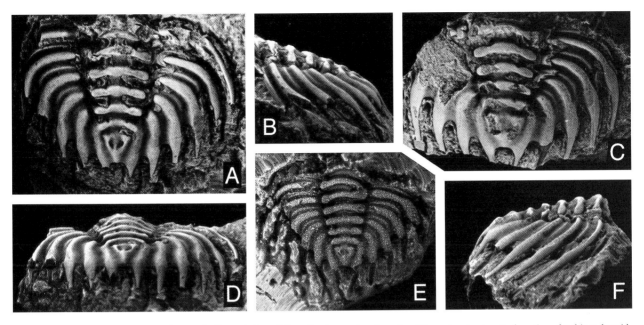

Fig. 82. Pliomeroides primigenus (Angelin, 1854). □A, B, D. Dorsal, lateral and posterior view of pygidium, showing terminal portion of rachis and pygidial spines; ×3, PMO 61483/1. Bjørkåsholmen, Asker. Coll. unknown. □C, F. Dorsal and lateral view of nearly complete pygidium; ×2.5, PMO 1393. Bjørkåsholmen, Asker. Coll. unknown, 1915. □E. Dorsal view of pygidium, showing pathological deformation of anterior rachial ring; ×4, PMO 33158. Nordre Grefsen at Gran, Hadeland. Coll.: T. Münster, 1893-08-31.

Emended description. – Sagittal length of cranidium one-third of posterior width. Glabella 20% longer (sag.) than wide (tr.), tapering slightly anteriorly. Occipital ring widest sagittally, tapering laterally behind L1 lobes. Occipital furrow slightly curved slightly crescentic anteriorly. Three pairs of deep, lateral glabellar furrows present, all reaching equally deep adrachially, directed slightly obliquely backwards at an angle of 10°. S1 furrows curving slightly backwards adrachially, almost isolating drop-shaped L1 lobes. S2 and S3 furrows curving less backwards adrachially (Fig. 81D), with equal distance between lobes. Anterior glabellar lobe rounded in front. Glabella lies in depression between fixed cheeks, outlined by well-defined dorsal furrow. Posterior fixed cheeks convex (sag.), narrow (sag.), elongated laterally, defined by a distinct ridge along glabella, pointed opposite to occipital furrow, sloping steeply down to dorsal furrow (Fig. 81H, J). Posterior margin curving slightly backwards, genal angles slightly pointed. Posterior border furrow widest (tr.) laterally, curving anteriorly to meet with almost transverse posterior facial suture. Palpebral lobes small (Fig. 81J), making up less than one-fifth of sagittal length, positioned away from glabella opposite to L2 lobes. Eye ridges converging anteriorly, meeting with dorsal furrow opposite S3 furrows. Anterior part of fixed cheeks nearly vertical; anterior facial sutures converging slightly anteriorly to meet nearly horizontal, brim-like, slightly undulating anterior border (Fig. 81I). Anterior border furrow distinct, deep sagittally and laterally, with deep pits oppo-

site anterolateral corners of glabella just anterior of eye ridges (Fig. 81D). Anterior margin semielliptical. Fixed cheeks have a reticulated structure, formed around deep pits (Fig. 81B, H), test on cheeks have small pits and granules. Free cheeks elongated with highly convex (tr.), subtriangular genal fields (Fig. 81C). Border furrow distinct, outlining slightly bevelled border. Structure on genal field similar to that of cranidium.

Table 34. Cranidial measurements of *Pliomeroides primigenus.*

Specimen	A	B	C	C1	C2	J	J1	J2	K	K1
1394	1.05	0.96	0.42	0.21	0.47	3.20	1.20	1.65	0.80	0.70
20055	0.95	0.90	1.25	0.02	0.12	2.46	1.08		0.63	0.56
36010	1.30	0.99	0.44	0.20	0.54	2.97	–	1.76	0.90	0.71
51212	1.30	–	–	–	–	3.58	–	–	–	–
72040/12	1.54	1.32	–	–	–	4.28	–	–	–	–
83828/1	1.28	1.21	0.60	0.20	0.49	3.68	1.68	2.00	0.90	0.78
83829/6	1.10	1.08	0.32	0.15	0.36	2.89	1.32	1.55	0.73	0.63
83830/1	0.90	0.82	–	–	0.35	2.12	1.18	1.46	0.64	0.60
83971/2	0.95	1.02	–	–	–	2.85	1.44	–	0.90	0.78
84030/1	1.22	1.20	0.50	0.26	0.64	4.36	1.73	2.24	1.00	0.90
84075/2	0.60	0.57	–	–	0.23	1.47	0.60	–	–	–
97155/1	1.15	1.09	0.40	0.23	0.53	3.25	1.30	1.60	0.89	0.69
97156/1	1.00	0.94	–	–	–	1.37	–	–	0.76	0.50
99190	0.90	0.85	0.29	0.16	0.32	2.28	1.06	1.19	0.63	0.53
117.055	0.65	0.57	0.18	0.08	0.23	1.19	–	–	0.45	0.41
121.539/1	1.38	1.20	0.43	0.34	0.46	4.05	1.72	2.20	1.12	1.05
121.614	1.13	0.90	–	–	–	3.04	–	–	0.72	–
121.615	0.75	0.66	0.24	0.90	0.35	1.64	–	–	0.48	0.32

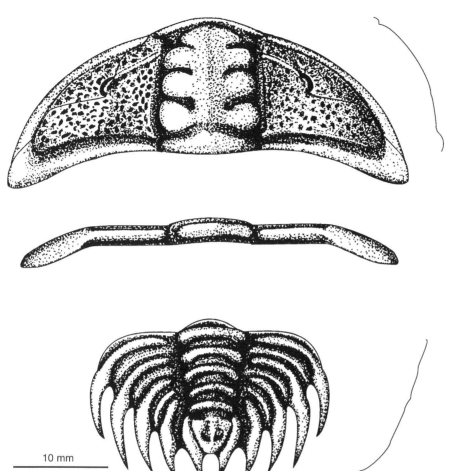

Fig. 83. Reconstruction of *Pliomeroides primigenus* showing the cephalon, a thoracic segment and the pygidium.

10 mm

Thoracic region with at least 11 segments (Fig. 81A). Rachial ring not well preserved, apparently having a narrow (sag.), convex (sag.) central ridge. Pleural fields narrow (exsag.), elongated (tr.). Convex (sag.) ridge along centre of pleurae becoming paddle-formed laterally, with slightly pointed termination. Narrow (sag.). anterior and posterior marginal fields attenuating laterally.

Sagittal length of pygidium somewhat less than half the maximum width. Anterior width of rachis one-third of maximum width (tr.), tapering posteriorly, somewhat pointed at posterior extremity (Fig. 82A), sloping almost vertically down to posterior margin (Fig. 82F). It has five convex (sag.), ridge-like rachial rings with wide (sag.) rachial furrows between. Terminal rachial lobe with paired depressions (Fig. 82A). Pleural field horizontal adrachially, sloping steeply laterally (Fig. 82D), nearly V-shaped in outline. Five pairs of pleural ridges curve outwards and rearwards, nearly exsagittally laterally, getting slightly wider at margin, then projecting into obliquely backward-directed, pointed spines (Fig. 82C, F).

Pleural furrows distinct, separated by thin interpleural ridges except between posterior pleurae (Fig. 82C). Outline semicircular.

Discussion. – The cranidium of *Pliomeroides primigenus* differs from that of the type species *P. deferrariisi* in having a proportionally wider glabella which tapers less markedly forward, having a posterior margin curving backwards laterally, lacking genal spines, a well-defined transition to dorsal furrow, a better-defined pattern on the fixed cheeks, and having a more prominent anterior margin. The pygidium differs in having a proportionally narrower and longer rachis with better-defined rachial

Table 35. Pygidial measurements of *Pliomeroides primigenus*.

Specimen	X	X1	Y	Z	W
1362/3	0.60	0.30	0.80	0.83	1.68
1365/1	0.74	0.38	1.37	1.40	2.10
1393	0.77	0.41	1.40	1.43	2.25
20110	0.33	0.18	0.65	0.68	1.11
33158	0.32	0.15	0.69	0.69	1.05
40753/1	0.60	0.30	1.05	1.32	2.10
61483/1	0.60	0.36	1.35	1.35	2.10
83949	0.42	0.18	0.72	0.72	1.20
121.563/3	0.50	0.26	0.74	0.75	1.50
121.594/4	0.45	0.24	0.66	0.66	1.26
121.613	0.45	0.21	0.74	0.75	1.20

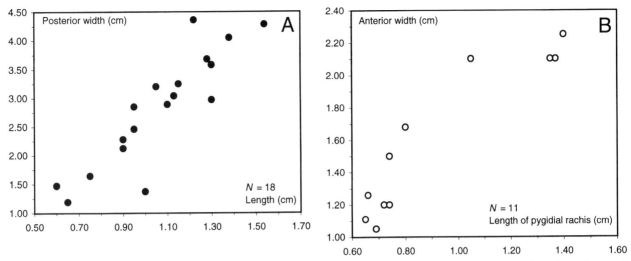

Fig. 84. Cranidia and pygidia of *Pliomeroides primigenus.* □A. Length (sag.) of cranidium plotted against posterior width (tr.) of cranidium. □B. Length (tr.) of rachis plotted against anterior width (tr.).

rings. The terminal rachial lobe is better-defined, with a pair of distinct pits, barely seen in the type species. Interpleural furrows are lacking only on the posterior part of the pleural field, while present only at the anterior part of the pleural fields in the type species. The pleural ridges extend laterally into more elongated and narrower pygidial spines in *P. primigenus.*

P. primigenus was formerly included in the genus *Evropeites* by Balashova (1966) and Fortey (1980), based on comparison between the type species *E. primigenus* (Angelin, 1854) var. *lamanskii* (Schmidt, 1907) and *E. hyperboreus* Fortey, 1980.

E. primigenus lamanskii is only known from the cranidium. It differs markedly in having a more rounded glabella with the S2 and S3 furrows isolated from the dorsal furrow, more backward-curving eyeridges, palpebral lobes situated opposite of 1P lobes, and less well-defined fixed cheeks adrachially.

The cranidium of *E. hyperboreus* Fortey, 1980, differs in having a slightly anteriorly expanding glabella, with the S3 furrows isolated from the dorsal furrow, eyeridges converging anteriorly at a lower angle to glabella, and a more rounded anterior margin parallel to the anterior part of glabella. The pygidium differs mainly in having broader pygidial ridges without pleural furrows. The species has at least twelve thoracic segments.

A small pygidium of *P. primigenus* (Fig. 82E) shows a teratological or pathological malformation where the right anterior pleural segment is not developed and the anterior rachial rings are fused on the right side. Owen (1985) and Babcock (1993) described and classified different abnormalities and malformations in trilobite morphologies.

Fig. 83 shows a reconstruction of *Pliomeroides primigenus,* and Fig. 84A, B shows length and width variables of the cranidia and pygidia, respectively.

Unidentified trilobite remains

Indet. sp. 1

Figs. 85A–C

Cephalon of late protaspid or early meraspid stage (PMO 121.628/2). Cephalon slightly wider than long, semicircular, convex, with narrow (tr.) glabella expanding anteriorly, reaching anterior margin. Occipital ring outlined. The affinity of this specimen is unknown. The expanding glabella could match with genera like *Euloma, Ceratopyge* or *Symphysurus.* These three genera are most common in the fauna, and growth series are known for all of them.

Indet. sp. 2

Figs. 85D

A doublure of an asaphid trilobite from a dark limestone nodule at the base of the formation. The anterior width of the specimen is about 1 cm (tr.). The terrace ridges are densely spaced on the recurved doublure. The rachis of the specimen is short. This may be a ptychopygine (R.A. Fortey, personal communication, 1996).

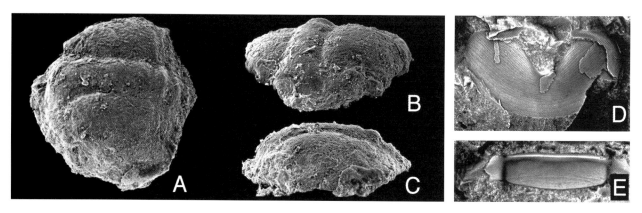

Fig. 85. Unidentified trilobite remains. □A–C. Obliquely dorsal, anterior and lateral view of protaspis stage of unknown species; ×100, PMO 121.628/2. Bjørkåsholmen, Asker. Coll.: F. Nikolaisen, 1960-05-01. □ D. Dorsal view of pygidial doublure, showing terrace ridges; ×4, PMO 84120. Found in dark limestone nodule 14–17 cm above base of the formation at Sjøstrand, Asker. Coll.: G. Henningsmoen, 1959-03-01. □E. Thoracic segment; ×5, PMO 1240/3. Found in dark limestone nodule 25–30 cm above base of the formation at Bjørkåsholmen, Asker. Coll. unknown, 1915.

Indet. sp. 3

Figs. 85E

Thoracic segment from a dark limestone nodule at the base of the formation. The specimen is about 1 cm (tr.). Wide (tr.), smooth rachial ring with very narrow rachial furrow. Thin transverse terrace ridges near posterior margin. Pleural fields subrectangular, posterolateral margin pointed. Pleural furrow diagonal. The affinity is difficult to establish. Thoracic segments of *Saltaspis stenolimbatus* n.sp. from the same level are unknown, but the pleurae on this segment are too short (tr.) to belong to this species, with its wide (tr.) posterior fixed cheeks.

References

Ahlberg, P. 1989a: Agnostid trilobites from the Lower Ordovician Komstad Limestone Formation of Killeröd, Scania, Sweden. *Palaeontology 32:3*, 553–570.

Ahlberg, P. 1989b: The type species of the Ordovician agnostid trilobite *Geragnostus* Howell, 1935. *Paläontologische Zeitschrift 63*, 309–317.

Ahlberg, P. 1992: Agnostid trilobites from the lower Ordovician of southern Sweden. *Transactions of the Royal Society of Edinburgh Earth Sciences 83*, 539–570.

Angelin, N.P. 1851: Palæontologia Suecica. Pars I. *Iconographia Crustaceorum Formationis Transitionis Fasc. I.* 24 pp. Weigel, Lund.

Angelin, N.P. 1854: Palæontologia Scandinavica. *Iconographia Crustacea Formationis Transitionis Fasc. II*, 21–92. Weigel, Lund.

Babcock, L.E. 1993: Trilobite malformations and the fossil record of behavioral asymmetry. *Journal of Paleontology 67:2*, 217–229.

Balashova, E.A. (Балашова Е.А.) 1960: Трилобиты среднего и верхнего ордовика и нижнего силура восточного Таймыра. [Middle and Upper Ordovician and Lower Silurian trilobites from eastern Taimyr.] *Izdatelstva Leningradskogo Universiteta 1960.* 111 pp. [In Russian.]

Balashova, E.A. (Балашова Е.А.) 1961: Некоторые тремадокские трилобиты Актюбинской области. [Some Tremadocian trilobites of the Aktyubinsk Province.] *Trudy Geologicheskogo Instituta, Akademiya Nauk SSSR 18*, 102–145. [In Russian.]

Balashova, E.A. (Балашова Е.А.) 1966: Трилобиты раннеордовиских отложений Русской платформы. [Early Ordovician trilobites from the Russian Platform.] *Voprosy Paleontologii 5*, 3–22. [In Russian.]

Balashova, E.A. 1974: Ontogeny of the trilobites *Ceratopyge forficula* and *Promegalaspides kasachstanensis. Paleontological Journal 4*, 486–492.

Barrande, J. 1846: *Notice préliminaire sur le Système Silurien et les Trilobites de Bohême I–IV.* 97 pp. Leipzig.

Barrande, J. 1852: Système Silurien du centre de la Bohême. Ière Partie. *Recherches Paléontologiques Vol. 1. Texte. Crustacés: Trilobites.* 935 pp.

Barton, D.C. 1915: A revision of the Cheirurinae, with notes on their evolution. *Washington University Study of Science Series 3:1*, 101–152.

Bassler, R.S. 1915: Bibliographic index of American Ordovician and Silurian fossils 1, 2. *United States Natural Museum Bulletin 92.* 1521 pp.

Berard, P. 1986: Trilobites de l'Ordovicien inférieur des Monts de Cabrières. *Memoires du Centre d'Etudes et de Recherches Géologiques et Hydrologiques 24.* 220 pp.

Bergeron, J. 1889: Etude géologique du massif ancien situé au Sud du Plateau Central. *Annales de Science Géologique 22 (Paris).* 362 pp.

Bergeron, J. 1895: Notes paléontologiques. *Bulletin de la Societé Géologique du France 23*, 465–484.

Beyrich, E. 1846: Untersuchungen über Trilobiten. Zweites Stück als Fortsetzung zu der Abhandlung „Ueber einige boehmische Trilobiten", 1–38. Reimer, Berlin.

Billings, E. 1861–1865: *Palaeozoic Fossils 1.* 426 pp. Geological Survey of Canada, Montreal.

Bjørlykke, K. 1974: Depositional history and geochemical composition of lower Palaeozoic epicontinental sediments from the Oslo Region. *Norges Geologiske Undersøkelse Bulletin 305.* 81 pp.

Bockelie, J.F. 1978: The Oslo Region during the Early Palaeozoic. *In* Ramberg, I.B. & Neumann, E.R. (eds.): *Tectonics and Geophysics of Continental Rifts*, 195–202. Reidel, Dordrecht.

Bockelie, J.F. & Nystuen, J.P. 1985: The south-eastern part of the Scandinavian Caledonides. *In* Gee, D.G. & Sturt, B.A. (eds.): *The Caledonide Orogen-Scandinavia and Related Areas*, 70–88. Wiley, Chichester.

Boeck, C.P.B. 1838: Uebersicht der bisher in Norwegen gefundenen Formen der Trilobiten-Familie. *In* Keilhau, B.M. (ed.): *Gæa Norvegica I*, 138–145. Johan Dahl, Christiania.

Bohlin, B. 1955: The Lower Ordovician Limestones between the *Ceratopyge* Shale and the *Platyurus* Limestone of Böda Hamn. *Bulletin of the Geological Institution of the University of Uppsala 35*, 111–151.

Bondarev, V.I, Burskij, A.Z., Koloskov, K.N. & Nekhorosheva, A.V. (Бондарев В.И., Бурский А.З., Колосков К.Н., Нехорошева А.В.) 1965: Раннеордовикская фауна юга Новой Земли и севера Пай-Хой и ее стратиграфическое значение. [Early Ordovician faunas from the south of Novaya Zemlja and the north of Phaj-Khoj and the stratigraphy.] *Nauchno-isslevdovatelskij Institut Geologii Arktiki, Ministerstva Geologii SSSR, Uchenye Zapiski, Paleontologiya i Biostratigrafiya 10, Leningrad*, 15–63. [In Russian.]

Bradley, J.H. 1925: Trilobites of the Beekmantown in the Philipsburg Region of Quebec. *Canadian Field Naturalist 39*, 5–9.

Branisa, L. 1965: Los fosiles, quaias de Bolivia. *Bolivia Service de Geologica 60*, 1–82.

Brøgger, W.C. 1882: Die silurischen Etagen 2 und 3 im Kristianiagebiet und auf Eker. *Universitätsprogramm für 2. Semester 1882, Kristiania.* 376 pp.

Brøgger, W.C. 1886: Über die Ausbildung des Hypostomes bei einigen skandinavischen Asaphiden. *Bihang till Kungliga Svenska Vetenskaps-Akademiens Handlingar 11:3.* 78 pp. (Also *Sveriges Geologiske Undersökning Serie C 82*).

Brøgger, W.C. 1896: Über die Verbreitung der Euloma–Niobe Fauna (der Ceratopygenkalk Fauna) in Europa. *Nyt Magazin for Naturvidenskab 36*, 164–240.

Bruton, D.L. 1968: The trilobite genus *Panderia* from the Ordovician of Scandinavia and the Baltic areas. *Norsk Geologisk Tidsskrift 48:1–2*, 1–53.

Bruton, D.L. & Harper, D.A.T. 1985: Early Ordovician (Arenig–Llanvirn) faunas from oceanic islands in the Appalachian–Caledonian Orogen. *In* Gee, D.G. & Sturt, B.A. (eds.): *The Caledonide Orogen – Scandinavia and Related Areas*, 359–368. Wiley, Chichester.

Bruton, D.L., Harper, D.A.T. & Repetski, J.E. 1989: Stratigraphy and faunas of the Parautochthon and Lower Allochthon of southern Norway. *In* Gayer, R.A. (ed.): *The Caledonide Geology of Scandinavia*, 231–241. Graham & Trotman, London.

Bruton, D.L., Lindström, M. & Owen, A.W. 1985: The Ordovician of Scandinavia. *In* Gee, D.G. & Sturt, B.A. (eds.): *The Caledonide Orogen – Scandinavia and Related Areas*, 237–282. Wiley, Chichester.

Bulman, O.M.B. & Rushton, A.W.A. 1973: Tremadoc faunas from boreholes in Central England. *Geological Survey of Great Britain Bulletin 43*, 1–40.

Burmeister, H. 1843: *Die Organisation der Trilobiten.* 148 pp. Reimer, Berlin.

Burskij, A.Z. (Бурский А.З.) 1966: Семейство Remopleurididae из ордовика севера Пай-Хойе, острова Вайгача и юга Новой Земли. [The family Remopleurididae from the Ordovician of northern Phaj-Khoj, Vajgach and southern Novaya Zemlja.] *Nauchno-isslvdovatelskij Institut Geologii Arktiki, Ministerstva Geologii SSSR, Uchenye Zapiski, Paleontologiya i Biostratigrafiya 12, Leningrad*, 30–45. [In Russian.]

Burskij, A.Z. (Бурский А.З.) 1970: Раннеордовикские трилобиты севера Пай-Хой. *In* [Bondarev, V.I.] Бондарев В.И. (ed.): Опорный разрез ордовика Пай-Хойе, Вайгача и юга Новой Земли. *[Ordovician key sections in Phaj-Khoj, Vajgach Island and southern Novaya Zemlja]*, 96–138 *Nauchno-issledovatelskij Institut Geologii Arktiki Ministerstva Geologii SSSR, Leningrad.* [In Russian.]

Callaway, C. 1877: On a new area of Upper Cambrian rocks in South Shropshire, with a description of a new fauna. *Quarterly Journal of the Geological Society of London 33*, 652–672.

Chang Went'ang & Fan Chiasung 1960: [Class Trilobita of the Ordovician and Silurian periods of the Chi-Lien Mountains.] *In* Yin Tsanhsun (ed.): *Geological Gazetteer of the Chi-Lien Mountains 4:1*, 83–148. Science Press, Beijing. [In Chinese.]

Chang Went'ang *et al.* 1964: *In: Atlas of Palaeozoic fossils from North Guizhou. Nanjing*, Pls. 5–10. [In Chinese.]

Chugaeva, M.N., Ivanova, V.A., Ordovskaya, M.M. & Yakovlev V.N. (Чугаева М.Н., Иванова В.А., Ордовская М.М. & Яковлев В.Н.) 1973: Биостратиграфия нижней части ордовика северо-востока СССР и биогеография ордовика. [Biostratigraphy of the lower part of the Ordovician in the north-east of the USSR and Ordovician biogeography.] *Akademiya Nauk SSSR. Ordena Trudovogo Krasnogo Znameni Geologicheskij Institut Trudy 213*. 303 pp. [In Russian.]

Chugaeva, M.N., Rozman, Kh.S. & Ivanova, V.A. (Чугаева М.Н., Розман Х.С. & Иванова В.А.) 1964: Сравнительная биострати-

графия ордовикских отложений северо-востока СССР. [Comparative biostratigraphy of Ordovician deposits in the north-east of the USSR.] *Akademiya Nauk SSSR, Geologicheskij Institut, Trudy 106.* 236 pp. [In Russian.]

Clark, T.H. 1924: The paleontology of the Beekmantown Series at Levis, Quebec. *Bulletins of American Paleontology 10:41*, 1–134.

Cooper, B.N. 1953: Trilobites from the Lower Champlainian formations of the Appalachian Valley. *Geological Society of America Memoirs 55*, 1–69.

Courtessole, R. & Pillet, J. 1975: Contribution à l'étude des faunes trilobitiques de l'Ordovicien inférieur de la Montagne Noire. Les *Eulominae* et les *Nileidae*. *Annales de la Société Géologique du Nord 95*, 251–272.

Courtessole, R. & Pillet, J. 1978: La faune des couches à *Shumardia* du Trémadocien supérieur de la Montagne Noire. *Société de Historique et Naturelle de Toulouse, Bulletin 114:1–2*, 176–186.

Crosfield, M.C. & Skeat, E.G. 1896: On the geology of the neighbourhood of Carmarthen. *Quarterly Journal of the Geological Society of London 52*, 523–541.

Dahll, T. 1857: Profil durch die Gegend von Skien, Porsgrunn und Langesund. *In* Kjerulf, T. 1857: Ueber die Geologie des südlichen Norwegens. *Nyt Magazin for Naturvidenskab 9*, 193–333.

Dalman, J.W. 1827: Om Palæaderna eller de så kallade Trilobiterna. *Kungliga Svenska Vetenskaps-Akademiens Handlingar 1827 (Separat, Stockholm 1828)*. 109 pp.

Dean, W.T. 1966: The Lower Ordovician stratigraphy and trilobites of the Landeyran Valley and the neighbouring districts of the Montagne Noir, south-western France. *Bulletin of the British Museum (Natural History) Geology 12:6*, 247–253.

Dean, W.T. 1973: The lower Palaeozoic stratigraphy and faunas of the Taurus Mountains near Beyşehir, Turkey, III. The trilobites of the Sobava Formation (Lower Ordovician). *Bulletin of the British Museum (Natural History) Geology 24:5*, 279–348.

Dean, W.T. 1989: Trilobites from the Survey Peak, Outram and Skoki Formations (Upper Cambrian – Lower Ordovician) at Wilcox Pass, Jasper National Park, Alberta. *Geological Survey of Canada Bulletin 389*. 141 pp.

Destombes, J., Sougy, J. & Willefert, S. 1969: Révision et découvertes paléontologiques (Brachiopodes, Trilobites et Graptolites) dans le Cambro-Ordovician du Zemmour (Mauritanie Septentrionale). *Bulletin de la Societé geologique de France 11*, 185–206.

Duan Ji-ye & An Su-lan 1986: Trilobites and cystoids. *In* Duan Ji-ye, An Su-lan & Zhao Da (eds.): The Cambrian–Ordovician boundary and its interval biotas, southern Lilin, Northeast China, 30–86. *J. Chanchun College of Geology, Special Issue on Stratigraphical palaeontology.*

Ebbestad, J.O.R. 1997: Bjørkåsholmen Formation (Upper Tremadoc) in Norway: regional correlation and trilobite distribution. *In* Stouge, S. (ed.): WOGOGOB-94 Symposium. *GEUS Rapport 1996/98*, 27–35.

Erdtmann, B.-D. 1965a: Eine spät-tremadocische Graptolitenfauna von Töyen in Oslo. *Norsk Geologisk Tidsskrift 45*, 97–112.

Erdtmann, B.-D. 1965b: Outline stratigraphy of graptolite-bearing 3b (Lower Ordovician) strata in the Oslo Region, Norway. *Norsk Geologisk Tidsskrift 45*, 481–547.

Erdtmann, B.-D. 1986: Early Ordovician eustatic cycles and their bearing on punctations in early nematophorid (planktic) graptolite evolution. *In* Walliser, O.H. (ed.): Global Bio-events, 139–152. *Lecture Notes in Earth Sciences 8.*

Erdtmann, B.-D. 1995: Tremadoc of East European Platform: stratigraphy, confacies regions, correlation and basin dynamics. *In* Cooper, J.D., Droser, M.L. & Finney, S.C. (eds.): *Ordovician Odyssey: Short Papers for the Seventh International Symposium on the Ordovician System. Las Vegas, Nevada, USA June 1995*, SEMP, Fullerton, California, USA, 237–239.

Erdtmann, B.-D. & Paalits, I. 1995: The Early Ordovician 'Ceratopyge Regressive Event' (CRE): its correlation and biotic dynamics across the East European Platform. *Lithuanian Geological Society, Geologija, 1994, 17*, 36–57.

Ergaliev, G.Kh. (Ергалиев Г.Х.) 1980: Трилобиты среднего и верхнего кембрия Малого Каратау. [Middle and Upper Cambrian trilobites of the Malyi Karatau Range.] 211 pp. *Akademiya Nauk Kazakhskoj SSR, Ordena Trudovogo Krasnogo Znameni Institut Geologicheskikh Nauk, Alma-Ata* [In Russian.]

[Fjelldal, Ø. 1966: The Ceratopyge limestone (3a) and limestone facies in the lower Didymograptus shale (3b) in the Oslo Region and adjacent districts. 129 pp. Unpublished Cand. Real. thesis, University of Oslo.]

Fortey, R.A. 1974: The Ordovician trilobites of Spitsbergen. I. Olenidae. *Norsk Polarinstitutt skrifter 160.* 129 pp.

Fortey, R.A. 1975: The Ordovician trilobites of Spitsbergen. II. Asaphidae, Nileidae, Raphiophoridae and Telephinidae of the Valhallfonna Formation. *Norsk Polarinstitutt skrifter 162.* 207 pp.

Fortey, R.A. 1980: The Ordovician trilobites of Spitsbergen. III. Remaining trilobites of the Valhallfonna Formation. *Norsk Polarinstitutt skrifter 171.* 163 pp.

Fortey, R.A. 1981: *Prospectatrix genatenta* (Stubblefield) and the trilobite superfamily Cyclopygacea. *Geological Magazine 118:6,* 603–614.

Fortey, R.A. 1983: Cambrian–Ordovician trilobites from the boundary beds in western Newfoundland and their phylogenetic significance. *In* Briggs, D.E.G. & Lane, P.D. (eds.): Trilobites and other early arthropods: papers in honour of Professor H.B. Whittington, F.R.S., 179–211. *Special Papers in Palaeontology 30.*

Fortey, R.A. 1986: The type species of the Ordovician trilobite *Symphysurus*: Systematics, functional morphology and terrace ridges. *Paläontologische Zeitschrift 60,* 255–275.

Fortey, R.A. & Chatterton, B.D.E. 1988: Classification of the trilobite suborder Asaphina. *Palaeontology 31:1 ,* 165–222.

Fortey, R.A. & Droser, M.L. 1996: Trilobites at the base of the Middle Ordovician, western United States. *Journal of Paleontology 70:1,* 73–99.

Fortey, R.A. & Owens, R.M. 1978: Early Ordovician (Arenig) stratigraphy and faunas of the Carmarthen district, south-west Wales. *Bulletin of the British Museum (Natural History) Geology 30,* 225–294.

Fortey, R.A. & Owens, R.M. 1987: The Arenig Series in South Wales. *Bulletin of the British Museum (Natural History) Geology 41:3,* 69–307.

Fortey, R.A. & Owens, R.M. 1991: A trilobite fauna from the highest Shineton Shales in Shropshire, and the correlation of the latest Tremadoc. *Geological Magazine 128,* 437–464.

Fortey, R.A. & Shergold, J.H. 1984: Early Ordovician trilobites, Nora Formation, Central Australia. *Palaeontology 27:2,* 315–366.

Fortey, R.A., Harper, D.A.T., Ingham, J.K., Owen, A.W. & Rushton, A.W.A. 1995: A revision of Ordovician series and stages from the historical type area. *Geological Magazine 132:1,* 15–30.

[Gjessing, J. 1976a: Tremadocian stratigraphy and fauna in the Oslo Region, Norway. 136 pp. Unpublished Cand. Real thesis, University of Oslo.]

Gjessing, J. 1976b: *Ceratopyge forficula* – et taksonomisk problem belyst ved hjelp av statistikk. *Fauna 29:3,* 134–141.

Goldfuss, A. 1843: Systematische Übersicht der Trilobiten und Beschreibung einiger neuer Arten derselben. *Neues Jahrbuch für Mineralogie, Geognostische und Geologische Petrefakten-Kunde 1843,* 537–567.

Hall, J. 1838: Description of two species of trilobites belonging to the genus *Paradoxides. American Journal of Science 33,* 139–142.

Hammann, W. 1974: Phacopina und Cheirurina (Trilobita) aus dem Ordovizium von Spanien. *Senckenbergiana lethaea 55:1/5,* 1–151.

Harper, D.T.A. & Owen A.W. 1983: The structure of the Ordovician rocks of the Ringerike district: evidence of a thrust system within the Oslo Region. *Norsk Geologisk Tidsskrift 61,* 111–115.

Harrington, H.J. 1937: On some Ordovician Fossils from Northern Argentina. *Geological Magazine 74,* 97–124.

Harrington, H.J. 1938: Sobre los faunas del Ordoviciano Inferior del Norte Argentino. *Revista del Museo de la Plata, Nueva Serie 1, Paleontología 4,* 109–289.

Harrington, H.J. & Kay, M. 1951: Cambrian and Ordovician faunas of eastern Colombia. *Journal of Paleontology 25,* 655–688.

Harrington, H.J. & Leanza, A.F. 1952: La clasificación de los Olenidae y los Ceratopygidae (Trilobita). *Revista Asociación Geologica Argentina 7,* 190–205.

Harrington, H.J. & Leanza, A.F. 1957: Ordovician trilobites of Argentina. *Department of Geology, University of Kansas, Special Publication 1.* 276 pp.

Hawle, I. & Corda A.J.C. 1847: Prodrom einer Monographie der böhmischen Trilobiten. *Königliche böhmischen Gesellschaft der Wissenschaften Abhandlung 5.* 176 pp.

Henderson, R.A. 1983: Early Ordovician faunas from the Mount Windsor Subprovince, northeastern Queensland. *In* Jell, P.A. & Roberts, J. (eds.): The Dorothy Hill Jubilee Memoir, 145–175. *Memoir of the Association of Australasian Palaeontologists 1.*

Henningsmoen, G. 1957: The trilobite family Olenidae. *Norsk Vitenskaps-Akademi Skrifter, Matematisk-naturvitenskapelig klasse I.* 304 pp.

Henningsmoen, G. 1958: The Upper Cambrian faunas of Norway. *Norsk Geologisk Tidsskrift 38,* 180–196.

Henningsmoen, G. 1959: Rare Tremadoc trilobites from Norway. *Norsk Geologisk Tidsskrift 39,* 153–173.

Henningsmoen, G. 1960: Cambro-Silurian deposits of the Oslo Region. *In* Holtedahl. N.O. (ed.): Geology of Norway. *Norges geologiske undersøkelse 208,* 130–150.

Hoel, O.A. 1999: Trilobites of the Hagastrand Member (Tøyen Formation, lowermost Arenig) from the Oslo Region, Norway. Part I: Asaphidae. *Norsk Geologisk Tidsskrift 79,* 79–103.

Hoel, O.A. (in press): Trilobites of the Hagastrand Member (Tøyen Formation, lowermost Arenig) from the Oslo Region, Norway. Part II: Remaining non-asaphid groups. *Norsk Geologisk Tidsskrift.*

Holliday, S. 1942: Ordovician trilobites from Nevada. *Journal of Paleontology 16,* 471–478.

Holm, G. 1897: Palæontologiske notiser 4. Om *Bohemilla(?) denticulata* Linrs. och *Remopleurides microphtalmus* Linrs. *Geologiska Föreningens i Stockholm Förhandlingar 19,* 457–482.

Holm, G. 1901: Kinnekulles berggrund. *Sveriges Geologiska Undersökning C 172,* 1–76.

Howell, B.F. 1935: Cambrian and Ordovician trilobites from Hérault, Southern France. *Journal of Paleontology 9:3,* 222–238.

Hsü, S.C. & Ma, C.T. 1948: The Ichangian Formation and the Ichangian fauna. *Contributions Natural Resources, Institute of Geology, Academica Sinica 8,* 1–51.

Hughes, C.P. 1979: The Ordovician trilobite fauna of the Builth Llandrindod Inlier. Wales. Part III. *Bulletin of the British Museum (Natural History) Geology 32:3,* 109–181.

Hughes, N.C. & Rushton, A.W.A. 1990: Computer-aided restoration of a Late Cambrian Ceratopygid trilobite from Wales, and its phylogenetic implications. *Palaeontology 33:2,* 429–445.

Hupé, P. 1955: Classification des trilobites. *Annales de Paléontologie (Paris) 39,* 91–325.

Hutchinson, R.D. 1952: The stratigraphy and trilobite faunas of the Cambrian sedimentary rocks of Cape Breton Island, Nova Scotia. *Geological Survey of Canada Memoirs 263.* 124 pp.

Ingham, J.K. & Tripp, R.P. 1991: The trilobite fauna of the Middle Ordovician Doulard Formation of the Girvan district, Scotland, and its palaeoenvironmental significance. *Transactions of the Royal Society of Edinburgh, Earth Sciences 82,* 27–54.

Jaanusson, V. 1956: On the trilobite genus *Celmus* Angelin, 1854. *Bulletin of the Geological Institution of the University of Uppsala 36,* 35–49.

Jaanusson, V. 1973: Aspects of carbonate sedimentation in the Ordovician of Baltoscandia. *Lethaia 6,* 11–34.

Jaanusson, V. 1976: Faunal dynamics in the Middle Ordovician (Viruan) of Baltoscandia. *In* Bassett, M.G. (ed.): The Ordovician System: proceedings of a Palaeontological Association Symposium, Birmingham, September 1974, 301–326. University of Wales Press and National Museum of Wales, Cardiff.

Jaanusson, V. 1979: Ecology and faunal dynamics. *In* Jaanusson, V., Laufeld, S. & Skoglund, R. (eds.): Lower Wenlock faunal and floral dynamics – Vattenfallet Section, Gotland. *Sveriges Geologiska Undersökning C 762*, 253–294.

Jaekel, O. 1909: Über die Agnostiden. *Zeitschrift der Deutsche Geologische Gesellschaft (Berlin)* 61, 380–401.

Jell, P.A. & Stait, B. 1985: Tremadoc trilobites from the Florentine Valley Formation, Tim Shea area, Tasmania. *Memoirs of the Museum of Victoria 46:1*, 1–34.

Jell, P.A., Hughes, N.C. & Brown, A.V. 1991: Late Cambrian (Post-Idamean) trilobites from the Higgins Creek area, western Tasmania. *Memoirs of the Queensland Museum 30:3*, 455–485.

Kjerulf, T. 1857: *Über die Geologie des Südlichen Norwegens.* 144 pp. Johan Dahl, Christiania.

Kjerulf, T. 1865: Veiviser ved geologiske excursioner i Christiania omegn. *Universitetsprogram for andet Halvaar 1865.* 43 pp. Brøgger & Christie's Bogtrykkeri, Christiania.

Klouček, C. 1931: *Orometopus* et autres fossiles nouveaux dans le dα² d'Olesná. *Věstník Státního Geologiceskoho Ústavu Československé republiky (Praha)* 7, 367–370.

Kobayashi, T. 1934: The Cambro-Ordovician formations and faunas of South Chosen. Palaeontology. Part II. Lower Ordovician faunas. *Journal of the Faculty of Science, Imperial University of Tokyo 3:9*, 521–585.

Kobayashi, T. 1936: On the Parabolinella fauna from the Province Jujuy, Argentina, with a note on the Olenidae. *Japanese Journal of Geology and Geography 13*, 85–102.

Kobayashi, T. 1953: On the Kainellidae. *Japanese Journal of Geology and Geography 23*, 37–61.

Kobayashi, T. 1955: The Ordovician fossils from the McKay Group in British Columbia, western Canada, with a note on the Early Ordovician palaeogeography. *Journal of the Faculty of Science, Imperial University of Tokyo 2:9:3*, 355–493.

Koken, E. 1896: Die Leitfossilien. 848 pp. Tauchnitz, Leipzig.

Kou, H.C., Duan, J.Y. & An, S.L. 1982: Cambrian–Ordovician boundary in the Northern China Platform with description of trilobites. *Paper for Fourth International Symposium on the Ordovician System*, 1–31.

Lake, P. 1906: Monograph on the British Cambrian Trilobites, Part I. *Palæontographical Society (London) Monographs 1907*, 1–28.

Lake, P. 1907: Monograph on the British Cambrian Trilobites, Part II. *Palæontographical Society (London) Monographs 1907*, 29–48.

Lake, P. 1908: Monograph on the British Cambrian Trilobites, Part III. *Palæontographical Society (London) Monographs 1908*, 49–64.

Lake, P. 1913: Monograph on the British Cambrian Trilobites, Part IV. *Palæontographical Society (London) Monographs 1912*, 65–88.

Lake, P. 1919: Monograph on the British Cambrian Trilobites, Part V. *Palæontographical Society (London) Monographs 1917*, 89–120.

Lake, P. 1931: Monograph on the British Cambrian Trilobites, Part VI. *Palæontographical Society (London) Monographs 1929*, 121–148.

Lake, P. 1932: Monograph on the British Cambrian Trilobites, Part VII. *Palæontographical Society (London) Monographs 1930*, 149–172.

Lake, P. 1940: Monograph on the British Cambrian Trilobites, Part XII. *Palæontographical Society (London) Monographs 1940*, 273–306.

Lake, P. 1942: Monograph on the British Cambrian Trilobites, Part XIII. *Palæontographical Society (London) Monographs 1942*, 307–332.

Lane, P.D. 1971: British Cheiruridae (Trilobita). *Palaeontographical Society (London) Monographs.* 95 pp.

Lane, P.D. & Thomas, A.T. 1983: A review of the trilobite suborder Scutelluina. *In* Briggs, D.E.G. & Lane, P.D. (eds.): Trilobites and other early arthropods: papers in Honour of Professor H.B. Whittington, F.R.S., 141–160. *Special Papers in Palaeontology 30.*

Lindholm, K. 1991: Ordovician graptolites from the Early Hunneberg of southern Scandinavia. *Palaeontology 34*, 283–327.

Lindström, G. 1888: *List of the Fossil Faunas of Sweden, Edited by the Palæontological Department of the Swedish State Museum (Natural History). I. Cambrian and Lower Silurian.* Stockholm. 29 pp.

Lindström, M. 1971: Vom Anfang, Hochstand und Ende eines Epikontinentalmeeres. *Geologische Rundschau 60*, 419–438.

Lindström, M. 1984: Baltoscandic conodont life environments in the Ordovician; sedimentologic and paleogeographic evidence. *Special Paper Geological Society of America 196*, 34–42.

Linnarsson, J.G.O. 1869: Om Västergötlands cambriska och siluriska aflagringar. *Kongliga Svenska Vetenskaps-Akademiens Förhandlingar 8*, 1–89.

Linnarsson, J.G.O. 1872: Anteckningar om den kambrisk-siluriska lagerserien i Jemtland. *Geologiska Föreningens i Stockholm Förhandlingar 1*, 34–47.

Linnarsson, J.G.O. 1873: Berättelse afgiven till Kongliga Vetenskaps-Akademien, om en med understöd af allmänna medel utförd vetenskaplig resa till Böhmen och Ryska Östersjö-provinserna. *Öfversigt af Kongliga Svenska Vetenskaps-Akademiens Förhandlingar 5*, 89–111.

Linnarsson, J.G.O. 1874: Försteningar från Lappland insamlade af Hrr. E. Sidenbladh och E. Erdmann. *Geologiska Föreningens i Stockholm Förhandlingar 2*, 129–131.

Linnarsson, J.G.O. 1875a: Anteckningar från en resa i Skånes silurtrakter 1874. *Geologiska Föreningens i Stockholm Förhandlingar 2*, 260–284.

Linnarsson, J.G.O. 1875b: Öfversigt af Nerikes öfvergångsbildningar. *Öfversigt af Kongliga Svenska Vetenskaps-Akademiens Förhandlingar 5*, 2–47.

Linnarsson, J.G.O. 1876: Geologiska iakttagelser under en resa på Öland. *Geologiska Föreningens i Stockholm Förhandlingar 3*, 71–86.

Linnarsson, J.G.O. 1878: On the trilobites of the Shineton shales. *Geological Magazine New Series 5.* 188 pp.

Linnarsson, J.G.O. 1879: Ceratopygekalk och undre graptolitskiffer på Falbygden. *Geologiska Föreningens i Stockholm Förhandlingar 4*, 269–270.

Lisogor, K.A. (Лисогор К.А.) 1954: Результаты монографической обработки фауны трилобитов нижнего силура Бет-Пак-Далы; Кендыктасних и Джебалинских гор. [Results of the monographic treatment of the Lower Silurian trilobite fauna of Bet-Pak-Daly, Kenduktask and Dzhebalinsk mountains.] *Geologya, Gornoye Delo, Metallurgiya. Kazahskij Gorno-Metallurgicheskij Institut 9*, 98 pp. [In Russian.]

Lisogor, K.A. (Лисогор К.А.) 1961: Трилобиты тремадокских и смежных с ними отложений Кендыктаса. [Tremadocian trilobites and their associates in the deposits of the Kendyktas.] *Trudy Geologicheskogo Instituta, Akademiya Nauk SSSR 18*, 55–91. [In Russian.]

Lisogor, K.A. (Лисогор К.А.) 1977: Биостратиграфия и трилобиты верхнего кембрия и тремадока Малого Каратау (южный Казахстан). [Biostratigraphy and trilobites of the Middle Cambrian and Lower Tremadoc in Malyj Karatau Range (southern Kazahstan).] *Institut Geologii i Geofiziki, Sibirskoe Otdelenie, Trudy 313*, 197–264. [In Russian.]

Lu Yanhao 1975: Ordovician trilobite faunas of central and Southwestern China. *Palaeontologia Sinica 152, New Series B 11*, 1–261 [in Chinese], 265–463 [In English].

Lu Yanhao & Lin Huanling 1984: Late Cambrian and earliest Ordovician trilobites of Jiangshan–Chanshan area, Zhejiang. *In*: Stratigraphy and Palaeontology of Systemic Boundaries in China. Cambrian–Ordovician boundary 1, 5–143. Anhui Science and Technology Publishing House, Heifei.

Ludvigsen, R. 1980: An unusual trilobite faunule from Llandeilo or Lowest Caradoc strata (Middle Ordovician) of northern Yukon Territory. *Current Research, Part B, Geological Survey of Canada, Paper 80-1B*, 97–106.

Ludvigsen, R. 1982: Upper Cambrian & Lower Ordovician trilobite biostratigraphy of the Rabbitkettle Formation, Western District of Mackenzie. *Life Sciences Contributions Royal Ontario Museum 134.* 188 pp.

Ludvigsen, R. & Tufnell, P.A. 1983: A revision of the Ordovician olenid trilobite *Triarthrus* Green. *Geological Magazine 120:6*, 567–577.

Löfgren, A. 1993: Conodonts from the lower Ordovician at Hunneberg, south-central Sweden. *Geological Magazine 130:2*, 215–232.

Männil, R. (Мянниль Р.) 1966: История развития Балтийского бассейна в ордовике. [The history of the evolution of the Baltic basin in the Ordovician.] *Eesti NSV teaduste Akadeemia. Geological Institute Tallinn*, 1–201. [In Russian.]

Mansuy, H. 1920: Nouvelle contribution à l'étude des faunes paléozoiques et mesozoiques de l'Annam septentrional, région de Thanh-Hoa. *Mémoires du Service géologique de l'Indochine 7:1*, 1–64.

Martinsson, A. 1974: The Cambrian of Norden. *In* Holland, C.H. (ed.): *Cambrium of the British Isles, Norden and Spitsbergen. Lower Palaeozoic Rocks of the World 2*, 185–283. Wiley, Chichester.

Matthew, G.F. 1893: Illustrations of the fauna of the St. John Group, No. VII. *Transactions of the Royal Society of Canada 10:4 for 1892*, 95–109.

M'Coy, F. 1846: *A synopsis of the Silurian fossils of Irland* (*Dublin*). 72 pp.

Mergl, M. 1984: Fauna of the Upper Tremadocian of Central Bohemia. *Sborník geologických věd. Paleontologie 26*, 9–46.

Mergl, M. 1994: Trilobite fauna from the Trenice Formation (Tremadoc) in Central Bohemia. *Folia Musei Rerum Naturalium Bohemia Occidentalis Seria Geologica 39*. 31 pp.

Moberg, J.C. 1890: Om en afdeling inom Ölands dictyonemaskiffer såsom motsvarighet till ceratopygeskifferen i Norge. *Sveriges Geologiska Undersökning C 109*, 1–9.

Moberg, J.C. 1900: Nya bidrag till utredning af frågan om gränsen mellan undersilur och kambrium. *Geologiska Föreningens i Stockholm Förhandlingar 22*, 523–540.

Moberg, J.C. & Möller, H. 1898: Om Acerocarezonen. *Geologiska Föreningens i Stockholm Förhandlingar 20*, 197–290.

Moberg, J.C. & Segerberg, C.O. 1906: Bidrag till kännedomen om Ceratopygeregionen med särskild hänsyn till dess utveckling i Fogelsongstrakten. *Meddelande Lunds Geologiska Fältklubb B 2*. 113 pp.

Modlinski, Z. 1973: Stratigraphy and development of the Ordovician in north-eastern Poland. *Instyitytut Geologiczny Prace LXXII* (*Warszawa*). 74 pp.

Moore, R.C. (ed.) 1959: Treatise on Invertebrate Paleontology. O. Arthropoda 1. *Geological Society of America and University of Kansas Press*. 560 pp.

Morris, S.F. & Fortey, R.A. 1985: *Catalogue of the Type and Figured Specimens of Trilobites in the British Museum (Natural History)*. 183 pp.

Nickelsen, R.P., Hossack, J.R., Garton, M., & Repetski, J. 1985: Late Precambrian to Ordovician stratigraphy and correlation in the Valdres and Synfjell thrust sheets of the Valdres area, southern Norwegian Caledonides, with some comments on sedimentation. *In* Gee, D.G. & Sturt, B.A. (eds.): *The Caledonide Orogene – Scandinavia and Related Areas*, 369–379. Wiley, Chichester.

Nielsen, A.T. 1995: Trilobite systematics, biostratigraphy and palaeoecology of the Lower Ordovician Komstad and Huk Formations, southern Scandinavia. *Fossils & Strata 38*. 374 pp.

Nikolaisen, F. 1991: The Ordovician trilobite genus *Robergia* Wiman, 1905 and some other species hitherto included. *Norsk Geologisk Tidsskrift 71*, 37–62.

Nikolaisen, F. & Henningsmoen, G. 1985: Upper Cambrian and Lower Tremadoc olenid trilobites from the Digermul peninsula, Finnmark, northern Norway. *Norges Geologiske Undersøkelse Bulletin 400*, 1–49.

Noltimer H.C. & Bergström S.M. 1977: Paleomagnetic evidence of the Baltic Shield and N. American continent during Early and Middle Ordovician time. *Third International Symposium on the Ordovician System, Columbus, Ohio 1977. Programs and Abstracts 5*, 501.

Nystuen, J.P. 1981: The late Precambrian 'Sparagmites' of southern Norway. A major Caledonian Autochthon – The Osen–Røa nappe complex. *American Journal of Science 281*, 69–94.

Nystuen, J.P. 1986 (ed.): Regler og råd for navnsetting av geologiske enheter i Norge. Av Norsk stratigrafisk komité. *Norsk Geologisk Tidsskrift 66 (Supplement 1)*. 96 pp.

Nystuen, J.P. 1989 (ed.): Rules and recommendations for naming geological units in Norway. *Norsk Geologisk Tidsskrift 69 (Supplement 2)*. 111 pp.

Owen, A.W. 1985: Trilobite abnormalities. *Transactions of the Royal Society of Edinburgh, Earth Science 67*, 255–272.

Owen, A.W., Bruton, D.L., Bockelie, J.F., Bockelie, T.G. 1990: The Ordovician successions of the Oslo Region, Norway. *Norges geologiske undersøkelse Special Publication 4*. 54 pp.

Owens, R.M., Fortey, R.A., Cope, J.C.W., Rushton, A.W.A. & Bassett, M.G. 1982: Tremadoc faunas from the Carmarthen district, South Wales. *Geological Magazine 119:1*, 1–112.

Pek, I. 1977: Agnostid trilobites of the Central Bohemian Ordovician. *Sborník Geologických věd. Paleontologie 19*, 7–44.

Peng, S. 1990: Tremadoc stratigraphy and trilobite faunas of northwestern Hunan. *Beringeria 2*. 170 pp.

Peng, S. 1992: Upper Cambrian biostratigraphy and trilobite faunas of the Cili–Taoyan area, northwestern Hunan, China. *Association of Australasian Palaeontologists Memoir 13*. 119 pp.

Perroud, H., Robardet, M. & Bruton, D.L. 1992: Palaeomagnetic constraints upon the palaeogeographic position of the Baltic Shield in the Ordovician. *Tectonophysics 201*, 97–120.

Petrunina, Z.E. (Петрунина З.Е.) 1973: Новые роды и виды тремадокских трилобитов западной Сибири. [New genera and species of Tremadoc trilobites in west Siberia.] *In* Seljatickij, G.A., Gurskaya, I.D., Yershov, P.V., Ivanov, I.G., Zapivalov, N.P., Miatkov, I.M., Rozhok, N.G., Skobyelev, Jo.D. & Staroverov, L.D. (Селятицкий Г.А., Гурская И.Д., Ершов П.В., Иванов И.Г., Запивалов Н.П., Мятков И.М., Рожок Н.Г., Скбылев Ю.Д. & Староверов Л.Д.) (eds.): Новые данные по геологии и полезным ископаемым западной Сибири. *Izdatel'stvo Tomskogo Universiteta 8*, Tomsk, 59–68. [In Russian.]

Petrunina, Z.E., Poletaeva, O.K., Semonova, V.S. & Fedyanina, E.S. (Петрунина З.Е., Полетаева О.К., Семенова В.С. & Федянина Е.С.) 1960: Тип Arthropoda [Phylum Arthropoda] *In* Khalfin, L.L. (Халфин Л.Л.) (ed.): Биостратиграфия палеозоя Саян-Алтайской горной области 1. Нижний палеозой. *Trudy Sibirskogo Nauchno-Issledovatel'skogo Instituta Geologii, Geofiziki i Mineral'nogo Syr'ya*, 409–433. (In Russian.)

Pillet, J. 1992: Le genre *Apatokephalus* Brögger 1896 (Remopleuridacea, Trilobite) dans l'Ordovicien inférieur de la Montagne Noir (Sud de la France). *Bulletin de la Société et science d'Anjou 14*, 23–33.

Pillet, J. & Courtessole, R. 1980: Révision de *Harpides* (*Dictyocephalites*) *villebruni* (Bergeron 1895) (Trilobite, Arénigien inférieur de la Montagne Noir, France méridionale). *Bulletin de la Société Géologique de France 22*, 414–420.

Pompeckj, J.F. 1902: Aus dem Tremadoc der Montagne Noire (Süd-Frankreich). *Neues Jahrbuch für Mineralogie, Geologie und Palaeontologie 1*, 1–8.

Popov, L. & Holmer, L.E. 1994: Cambrian–Ordovician lingulate brachiopods from Scandinavia, Kazakhstan, and South Ural Mountains. *Fossils & Strata 35*. 156 pp.

Post, L. von: 1906: Bidrag till kännedomen om Ceratopygeregionens utbildning inom Falbygden. *Geologiska Föreningens i Stockholm Förhandlingar 28:6*, 465–480.

Poulsen, V. 1965: An early Ordovician trilobite fauna from Bornholm. *Meddelelser fra Dansk Geologisk Forening 16:1*. 113 pp.

Přibyl, A. & Vaněk, J. 1980: Ordovician trilobites of Bolivia. *Rozpravy Československé Akademie věd 90:2*, 3–90.

Qian Yiyuan 1985: Late Cambrian trilobites from the Tangcun Formation of Jingxian, southern Anhui. *Palaeontologia Cathayana 2*, 137–167. [In Chinese.]

Rasetti, F. 1943: New Lower Ordovician trilobites from Quebec. *Journal of Paleontology 17*, 101–104.

Rasetti, F. 1954: Early Ordovician trilobites faunules from Quebec and Newfoundland. *Journal of Paleontology 28*, 581–587.

Raw, F. 1949: Facial sutures in the (hypoparian) trilobites *Loganopeltoides* and *Loganopeltis*, and the validity of these genera. *Journal of Paleontology 23*, 510–514.

Raymond, P.E. 1913: Some changes in the names of genera of trilobites. *Ottawa Naturalist (Ottawa) 26:11*, 137–142.

Raymond, P.E. 1922: The Ceratopyge fauna in western North America. *American Journal of Science III 15*, 204–210.

Raymond, P.E. 1924: New Upper Cambrian and Lower Ordovician trilobites from Vermont. *Proceedings of the Boston Society of Natural History 37*, 389–466.

Raymond, P.E. 1925: Some trilobites of the Lower Middle Ordovician of eastern North America. *Bulletin of the Museum of Comparative Zoology, Harvard University 64:2*, 273–296.

Raymond, P.E. 1937: Upper Cambrian and Lower Ordovician Trilobita and Ostracoda from Vermont. *Geological Society of America Bulletin 48*, 1079–1146.

Reed, F.R. 1903: The Lower Palaeozoic trilobites of the Girvan district, Ayrshire, I. *Palæontographical Society (London) Monographs*, 1–48.

Reed, F.R. 1931: A review of the British species of Asapidae, II. *Annual Magazine of Natural History 7:10*, 441–472.

Regnéll, G. 1940: Note on *Ceratopyge forficula* (Sars). *Kungliga Fysiografiska Sällskapets i Lund Förhandlingar 9:8*, 1–3.

[Ribland-Nilssen, I. 1985: Kartleggning av Langesundshalvøyas Kambro-ordoviciske avsetningslagrekke, intrusiver og forkastnings-tektonikk, samt fullført lithostratigrafisk inndeling av områdets mellom-Ordovicium. Unpublished Cand. Scient. thesis, University of Oslo. 176 pp.]

Robison, R.A. & Pantoja-Alor, J. 1968: Tremadoc trilobites from the Nochixtlán Region, Oaxaca, Mexico. *Journal of Paleontology 42:3*, 767–800.

Ross, R.J. 1951: Stratigraphy of the Garden City Formation in northeastern Utah, and its trilobite faunas. *Peabody Museum of Natural History, Yale University (New Haven) Bulletin 6*, 1–161.

Ross, R.J. 1958: Trilobites in a pillow-lava of the Ordovician Valmey Formation, Nevada. *Journal of Paleontology 32:3*, 559–570.

Ross, R.J. 1970: Ordovician brachiopods, trilobites and stratigraphy in eastern and central Nevada. *Geological Survey Professional Paper 639*. 103 pp.

Rozova, A.V. (Розова А.В.) 1960: Верхнекембрийские трилобиты Салаира (толсточихинская свита). [Upper Cambrian trilobites of Salair (Tolstochikhin series).] *Akademiya Nauk SSSR, Sibirskoe Otdelenie, Trudy Instituta Geologii i Geofiziki 5*, 94 pp. [In Russian.]

Rozova, A.V. (Розова А.В.) 1963: Биостратиграфическая схема расчленения верчнего и верхов среднего кембрия северо-запада Сибирской платформы и новые верхнекембрийские трилобиты р. Кулюмбе. [Biostratigraphic scheme for subdividing the upper part of Middle Cambrian of northwestern Siberian platform and new Upper Cambrian trilobites of the River Kulyumbe area.] *Geologiya i geofizika 9*, 3–19. [In Russian.]

Rozova, A.V. (Розова А.В.) 1968: Биостратиграфия и трилобиты верхнего кембрия и нижнего ордовика северо-запада Сибирской платформы. [Biostratigraphy and trilobites of the Upper Cambrian and Lower Ordovician of the northwestern Siberian Platform.] *Akademiya Nauk SSSR, Sibirskoe Otdelenie, Trudy Instituta Geologii i Geofizki 36*. 243 pp. [In Russian.]

Rozova, A.V. (Розова А.В.) 1983: Ремоплевридиоиды и протапатокефалоиды (трилобиты). [Remopleurids and Proapatokephaloids (Trilobita).] *In* Dagys, A.S. & Dubatolov, V.N. (Дагыс А.С. & Дубатолов В.Н.) (eds.): Морфология и систематика беспозвоночных фанерозоя. [Morphology and systematics of Phanerozoic invertebrates.] *Akademiya Nauk SSSR, Sibirskoe Otdelenie Trudy Instituta Geologii i Geofizki 538*, 127–138. [In Russian.]

[Rudolf, F. 1992: Kopfmuskulatur bei Trilobiten. Rekonstruktion, Funktionsmorphologie und phylogenetisch-systematische Schlußfolgerung. Unpublished Ph. D. thesis, Zoologischen Institut der Christian-Albrechts-Universität Kiel. 116 pp.]

Rushton, A.W.A. 1982: The biostratigraphy and correlation of the Merioneth-Tremadoc Series boundary in North Wales. *In* Bassett, M.G. & Dean, W.T. (eds.): The Cambrian-Ordovician boundary: sections, fossil distributions, and correlations. *National Museum of Wales, Geological Series 3*, 41–59.

Rushton, A.W.A. 1988: Tremadoc trilobites from the Skiddaw Group in the English Lake District. *Palaeontology 31*, 677–698.

Růžička, K. 1926: Fauna vrstev Eulomových rudního ložiska u Holoubkova (v Ouzkém). Část I. Trilobiti. [The fauna of the Euloma beds from a quarry near Holoubkova (in Ouzk). Part I. Trilobites.] *Rozprozi Československé Akademii Věd Umění II 35:39*, 1–26. [In Czech.]

Růžička, K. 1934: Příspěvek k poznání trilobitů Barrandienu. [Contribution to the knowledge of the trilobites of the Barrandien.] *Rozprozi Československé Akademii Věd Umění II 44:11*, 1–10. [In Czech.]

Salter, J.W. 1849: Figures and descriptions illustrative of British organic remains. *Memoirs of the Geological Survey of Great Britain, Decade 2*. 39 pp.

Salter, J.W. 1864–1883: A monograph of the British trilobites. *Palæontographical Society (London) Monographs*. 244 pp.

Salter, J.W. 1866: The geology of north Wales. *Geological Survey of Great Britain 3*, 239–381.

Sars, M. 1835: Über einige neue oder unvollständig bekannte Trilobiten. *Oken Isis (or Encyclopedische Zeitung) 1835*, 333–343.

Schmidt, F. 1907: Revision der Ostbaltischen Silurischen Trilobiten. Abt. 6. *Mémoire de l'Académie Imperiale des Science de St. Petersburg 8:20:6*, 1–93.

Schrank, E. 1972: *Nileus*-Arten (Trilobita) aus Geschieben des Tremadoc bis tieferen Caradoc. *Der deutsche Gesellschaft für geologische Wissenschaft A. Geologie und Paläeontologie 17:3*, 351–375.

Schrank, E. 1973: *Nileus exarmatus glazialis* nom. nov. pro *N. e. lineatus* Schrank 1972. *Zeitschrift für geologische Wissenschaft 1*, 1186.

Sdzuy, K. 1955: Die Fauna der Leimitz-Scheifer (Tremadoc). *Abhandlungen der Senckenbergischen Naturforschenden Gesellschaft 492*, 1–74.

Sdzuy, K. 1958: Fossilien aus dem Tremadoc der Montagne Noire. *Senckenbergiana lethaea 39*, 255–285.

Shaw, A.B. 1951: The paleontology of northwestern Vermont. I. New late Cambrian trilobites. *Journal of Paleontology 25*, 97–114.

Shaw, A.B. 1957: Quantitative trilobite studies II. Measurements of the dorsal shield of non-agnostidean trilobites. *Journal of Paleontology 31*, 193–207.

Shergold, J.H. 1980: Late Cambrian trilobites from the Chatsworth Limestone, western Queensland. *Bulletin of the Bureau of Mineral Resources, Geology and Geophysics 186*. 111 pp.

Shergold, J.H. & Sdzuy, K. 1984: Late Cambrian and early Tremadoc trilobites from Sultan Dag, central Turkey. *Senckenbergiana lethaea 65*, 51–135.

Shergold, J.H., Laurie, J.R. & Sun Xiaowen 1990: Classification and review of the trilobite order Agnostida Salter, 1864: an Australian perspective. *Report of the Bureau of Mineral Resources, Geology and Geophysics 296*. 93 pp.

Skjeseth, S. 1952: On the Lower Didymograptus zone (3b) at Ringsaker, and contemporaneous deposits in Scandinavia. *Norsk Geologisk Tidsskrift 30*, 138–182.

Speyer, S.E. 1987: Comparative taphonomy and palaeocology of trilobite lagerstätten. *Alcheringa 11*, 205–232.

Spjeldnæs, N. 1961: Ordovician climatic zones. *Norsk Geologisk Tidsskrift 41*, 45–57.

Størmer, L. 1920: Om nogen fossilfund fra etage 3aα ved Vækerø, Kristiania. *Norsk Geologisk Tidsskrift 4*, 1–15.

Størmer, L. 1940: Early descriptions of Norwegian trilobites. *Norsk Geologisk Tidsskrift 20*, 113–151.

Størmer, L. 1967: Some aspects of Caledonian geosyncline and foreland west of the Baltic Shield. *Quarterly Journal of the Geological Society of London 123*, 183–214.

Stubblefield, C.J. 1926: Notes on the development of a trilobite, *Shumardia pusilla* (Sars). *Journal of the Linnean Society (Zoology) 36*, 345–372.

Stubblefield, C.J. & Bulman, O.M.B. 1927: The Shineton Shales of the Wrekin district. *Quarterly Journal of the Geological Society of London 83*, 96–145.

Thomas, A.T., Owens, R.M. & Rushton, A.W.A. 1984: Trilobites in British stratigraphy. *Special Report of the Geological Society of London 16*. 78 pp.

Thoral, M.M. 1935: Contribution à l'Ordovicien inférieur de la Montagne Noir et révision sommaire de la faune Cambrienne. *Imprim. La Charité, Montpellier*. 362 pp.

Thoral, M.M. 1946: Cycles géologiques et formations nodulières de la Montagne Noire. *Nouvelles Archives du Muséum d'Historie Naturelle de Lyon. Fascicule I*. 3–99.

Tjernvik, T.E. 1955: *Nericiaspis*, a new genus of proparian olenids. *Geologiska Föreningens i Stockholm Förhandlingar 77:2*, 209–212.

Tjernvik, T.E. 1956a: On the Early Ordovician of Sweden. *Bulletin of the Geological Institutions of Uppsala 35*, 108–284.

Tjernvik, T.E. 1956b: The Ceratopyge Region. *In* Pruvost, P. (ed.): *Lexique Stratigraphique International Vol. I Europa*, 61.

Tjernvik, T.E. & Johansson, J.V. 1980: Description of the upper portion of the drill-core from Finngrundet in the South Bothnian Bay. *Bulletin of the Geological Institution of the University of Uppsala. New Series 8*, 173–204.

Torsvik, T.H., Ryan, P.O., Trench A. & Harper, D.A.T. 1990: Cambrian–Ordovician palaeogeography of Baltica. *Geology 19*, 7–10.

Turner F.E. 1940: *Alsataspis bakeri*, a new Lower Ordovician trilobite. *Journal of Paleontology 14*, 516–518.

Ulrich, E.O. & Resser, C.E. 1930: The Cambrian of the Upper Mississippi Valley. Part 1, Trilobita; Dikelocephalinae and Osceolinae. *Bulletin of the Public Museum of the City of Milwaukee 12*, 1–122.

Vaněk, J. 1965: Die Trilobiten des mittelböhmischen Tremadoc. *Senckenbergiana lethaea 46*, 263–308.

Vodges, A.W. 1890: A bibliography of Palaeozoic Crustacea from 1698 to 1889, including a list of North American species and a systematic arrangement of genera. *U.S. Geological Survey Bulletin 63*. 177 pp.

Vogdes, A.W. 1925: Paleozoic Crustacea. Parts I–III. *Transactions of the San Diego Society of Natural History 4*, 1–154.

Walcott, C.D. 1884: Paleontology of the Eureka District. *U.S. Geological Survey Memoire 8*, 1–286.

Walcott, C.D. 1914: *Dikelocephalus* and other genera of the Dikelocephalinae. *Smithsonian Miscellaneous Collections 57:13*, 345–412.

Wallerius, J.D. 1895: Undersökninger öfver zonen med *Agnostus laevigatus* i Västergötland. *Academiska Afhandlingar Lund 1895*. 71 pp.

Weber, V.N. (Вебер В.Н.) 1932: Трилобиты Туркестана. [Trilobites of the Turkestan.] *Trudy vsesoyuznoj geologo-razvedochnogo ob'edineniya, N.S. 178*, 3–151. [In Russian with English summary.]

Westergård, A.H. 1909: Studier öfver Dictyograptusskiffern och dess gränslager med särskilt hensyn till i Skåne förekommande bildningar. *Lunds Universitets Årsskrift N.F. Afdeling 2, 5:3*, and also *Kungliga Fysiografiska Sällskapets Handlingar. N. F. 20:3*, 1–79.

Westergård, A.H. 1939: On Swedish Cambrian Asaphidae. *Sveriges Geologiska Undersökning C 421*, 1–16.

Whittington, H.B. 1950: Swedish Lower Ordovician Harpidae and the genus *Harpides*. *Geologiska Föreningens i Stockholm Förhandlingar 72*, 301–306.

Whittington, H.B. 1965: Trilobites of the Ordovician Table Head Formation, western Newfoundland. *Bulletin of the Museum of Comparative Zoology 132:4*, 275–442.

Whittington, H.B. & Almond, J.E. 1987: Appendages of the Ordovician trilobite *Triarthrus eatoni*. *Philosophical Transactions of the Royal Society of London 317*, 1–46.

Whittington, H.B. & Hughes, C.P. 1974: Geography and faunal provinces in the Tremadoc epoch. *In* Ross, C.A. (ed.): *Paleogeographic Provinces and Provinciality. Society of Economic Paleontologists and Mineralogists Special Publication 21*, 203–218.

Wilson, J.L. 1954: Late Cambrian and Early Ordovician trilobites from the Marathon Uplift, Texas. *Journal of Paleontology 28*, 249–258.

Wilson, J.L. 1975: *Carbonate Facies in Geological History*. 471 pp. Springer, Berlin.

Wiman, C. 1905: Ein Shumardiaschifer bei Lanna in Nerike. *Arkiv för Zoologi 2:11*, 1–20.

Wiman, C. 1907: Studien über das Nordbaltische Silurgebiet II. *Bulletin of the Geological Institution of Uppsala 8*, 6–168.

Wolfart, R. 1970: Fauna, Stratigraphie und Paläogeographie des Ordoviziums in Afghanistan. *Beiheft zum Geologischen Jahrbuch 89*. 125 pp.

[Wöltje, J. 1989: Zur Geologie im Modum/Øvre Eiker Gebit am Westrand des Oslograbens, Südnorwegen, unter besonderer Berücksichtigung der Bjerkåsholmen-Formation, oberes Tremadoc. Unpublished Diplom thesis, Georg-August-Universität, Göttingen. 87 pp.]

Xiang Li-wen & Zhang Tai-Rong 1984: Tremadocian trilobites from the western part of northern Tanshan, Xinjiang. *Acta Palaeontologica Sinica 23:4*, 399–410. [In Chinese with English abstract.]

Yin Gongzeng & Li Shanji 1978: [Class Trilobita.] *In [Palaeontological Atlas of Southwest China. Fasc. 1, Guizhou Province. 1, Cambrian to Devonian periods.]* Geological Publishing House, Beijing, 385–595. [In Chinese.]

Zhou Tien-Mei, Liu Li-Jeng, Meng Xiansong & Sun Zheng-Hua 1977: [Class Trilobita.] *In* Wang Xiao-Feng & jin Yu-Qin (eds.): *[Palaeontological Atlas of Central and South China. I. Early Palaeozoic.]* Geological Publishing House, Beijing, 104–266. [In Chinese.]

Ziegler, P.A. 1982: *Geological Atlas of Western and Central Europe*. 130 pp. Elsevier, Amsterdam.